The Gamefulness of American Postmodernism

Studies in Literary Criticism and Theory

Hans H. Rudnick
General Editor

Vol. 10

PETER LANG
New York • Washington, D.C./Baltimore • Boston • Bern
Frankfurt am Main • Berlin • Brussels • Vienna • Oxford

Steven D. Scott

The Gamefulness of American Postmodernism

John Barth & Louise Erdrich

PETER LANG
New York • Washington, D.C./Baltimore • Boston • Bern
Frankfurt am Main • Berlin • Brussels • Vienna • Oxford

Library of Congress Cataloging-in-Publication Data

Scott, Steven D.
The gamefulness of American postmodernism:
John Barth and Louise Erdrich / Steven D. Scott.
p. cm. — (Studies in literary criticism and theory; vol. 10)
Includes bibliographical references (p.) and index.
1. Barth, John—Criticism and interpretation. 2. Erdrich, Louise—Criticism and interpretation. 3. American fiction—20th century—History and criticism.
4. Postmodernism (Literature)—United States. 5. Game theory. I. Title.
II. Studies in literary criticism and theory; v. 10.
PS3552.A75Z517 813'.5409113—dc21 99-043009
ISBN 0-8204-3894-4
ISSN 1073-2004

Die Deutsche Bibliothek-CIP-Einheitsaufnahme

Scott, Steven D.:
The gamefulness of American postmodernism:
John Barth and Louise Erdrich / Steven D. Scott.
–New York; Washington, D.C./Baltimore; Boston; Bern;
Frankfurt am Main; Berlin; Brussels; Vienna; Oxford: Lang.
(Studies in literary criticism and theory; Vol. 10)
ISBN 0-8204-3894-4

The paper in this book meets the guidelines for permanence and durability
of the Committee on Production Guidelines for Book Longevity
of the Council of Library Resources.

Printed in the United States of America

I would really like to have slipped imperceptibly into this.... I would have preferred to be enveloped in words, borne away beyond all possible beginnings. At the moment of speaking, I would like to have perceived a nameless voice, long preceding me, leaving me merely to enmesh myself in it, taking up its cadence, and to lodge myself, when no one was looking, in its interstices as if it had paused an instant, in suspense, to beckon to me. There would have been no beginnings: instead, speech would proceed from me, while I stood in its path—a slender gap—the point of its possible disappearance. (Foucault, "The Discourse on Language" 215)

TABLE OF CONTENTS

CHAPTER ONE

Introduction

I begin this study with several assumptions which comprise a context for what follows: first, that postmodernist literature tends to be what Roland Barthes has called "writerly" literature—that is, it tends to be composed of those texts in which the reader is "no longer a consumer, but a producer of a text" (*S/Z* 4); second, that reading is (therefore) an active process (Iser, *The Act of Reading*); and third, that "meaning" is that which is produced in the interactions among readers, texts, and contexts, in the process that is called reading (Fish, *Is There a Text in This Class?*).

One consequence of these assumptions is that postmodernism as it is conceived in this study is at least in part open-ended, impermanent, and changeable—it is playful. Mihai Spariosu notes that "literature ... has from the very beginning been associated with play" (*Dionysus* 26). Theorists of the postmodern have expanded the "association" of play and literature into a constitutive part of literary postmodernism. Ruth Burke, in her recent book on postmodern criticism and fiction, *The Games of Poetics*, writes that play "is to be understood as a human and humanizing activity" (1); and further, that "play is an attitude" (34), one that appears insistently, and necessarily, in postmodernism. Brian Edwards argues in his dissertation, "The Player Played," that "[postmodernist] literature is ... playful in its ability to transcend, by mockery, parody, or experiment, its own `rules' or conventions" and that, furthermore, "[postmodernist] literature is unavoidably ambiguous by virtue of its allusiveness and its limited control over reader activity" (16).[1] Edwards's argument is that play is an essential component of literary postmodernism, in part because "play denotes a perspective ... which affirms freedom and possibility against restriction, resignation and closure" (32). Burke and Edwards and many others have argued that part of what makes postmodernism distinctive, then, is precisely its playful attitude and perspective. Against that playfully changeable model stand the socio-economic theories of Fredric Jameson, for example, wherein "late capitalist" market pressures produce a model of postmodernism that is centred not on process, attitude, openness—play—but rather on production and consumption: product, work, economics. The two positions are apparently irreconcilable, although they describe many of the same phenomena, and sometimes in similar terms.

I shall argue that it is precisely in the tensions that are created between these readings and models that postmodernist literature begins to make its shadowy presence felt. Further, neither the model based on play that Burke and Edwards and others construct, nor the model based on economic pressures that Jameson (and others) sees as so definitively shaping the postmodern, can adequately map literary postmodernism. In fact, I think that it is in the tensions between the two that postmodernist literature arises, not so much in the gap between models as in the intersections and gaps between attitudes. Further, I think that a model that would usefully accommodate both camps is a model that would construct postmodernist literature as fundamentally not only playful, but gamelike. That model would wed notions of play to notions of economic rules and constraints, and produce a gamelike structure within which literary postmodernism can be discussed.

It is with an attitude of open playfulness that the reader of the postmodernist text must come to her task; but that playful attitude will always be tempered in the first place by the knowledge that the text in hand is also a product: it was, after all, packaged and marketed, and something that was more, or less, expensive to buy. The reader of the postmodernist novel must also not only realize that the novel is by tradition a closed, finished, well-crafted product of artistic creation, but that postmodernism constructs itself as an ongoing process and as a project that resists closure; the postmodernist novel is always, therefore, both a carefully polished linguistic construct, on one hand, and the site of open-ended play, on the other: the postmodernist novel is always the site of fundamental paradoxes and tensions.

The first section of this book is composed of two chapters. The first of those chapters is a discussion of some of the myriad theories of what postmodernism is, what it does and what it can do, where it is located, and how one might go about finding or constructing it. The second chapter raises questions of play, game, conventions, and rules, and suggests a model that I hope will begin to make sense of some of the more slippery areas that together constitute literary postmodernism. My model suggests specific contextual limits and frames—rules—that not only delimit the play of meanings, but that, in fact, make meaningful playful activity possible.

The second section of this book is composed of readings from within that model of recent novels written by two significant postmodernist writers, each of whom attacks the problem of inscribing the postmodern in ways very different from the other's: that the two of them appear together here, in fact, is itself probably an appeal to Ludwig Wittgenstein's theories of "family resemblances."

I want to argue in this book that, although postmodernist literature is playful, the concept of play alone does not do justice to the postmodernist literary enterprise. Meanings may be slippery; they may be finally undecidable; they may play endlessly; they may be, in fact, as Jacques Derrida claims, nothing but the endless free play of "différance." Play *is* open-ended, undecided, and imprecise, as is postmodernism itself; but there are within particular expressions of postmodernism frames, limits, and contexts at work that allow the label "novel," for instance, to continue to make sense. "Playful" alone is not enough to describe what a postmodernist novel is or does, for such novels are open, playful, and resistant to closure, but are also, and at the same time, polished and highly crafted works of art.

Edwards establishes very well the importance of play to postmodernism; however, more than "play" is needed to make sense of literary postmodernism—it is necessary to provide a context for that play. In this study, I shall investigate what it means for postmodernism to be game-like. I shall claim that the concepts of games and gamefulness are important to discussions of the literary postmodern, in part because what we have otherwise is without form, structure, or meaning (however provisional and changeable). Thus my model will balance the mutability of play and postmodern playfulness with the apparent rigidity of games, rules, limits, and borders. It is, therefore, not true, as Edwards claims, that game is "subsumed" by play (18); play does not take the place of game; it does not drive game out. Instead, game structures play, and thereby makes play itself possible. Without the sort of structure that is provided to play by game, meaning, however provisional and fleeting it remains, becomes impossible.

Finally, while I think that notions of games, of gamefulness, and of game-like structures are important in discussions of the postmodern, I shall not claim in this book that literature itself *is* a game. My claim here is merely that game structures and notions of gamefulness are both useful and important in discussions of literary postmodernism, and that the postmodern novel structures itself in game-like ways. In fact, games and gamefulness have a function here which is much like the function of the frame in aesthetics. Susan Stewart writes in *Nonsense*:

> It is an old argument in aesthetics that it is not the picture that makes art art, but the frame and the frame's implicit message "This is art." The frame focuses our attention not upon content alone, but upon the organization of content and the relationship between content and its surroundings. The idea of content itself is brought about by organizing interpretive activities. (21)

Robert Wilson argues in his book *In Palamedes' Shadow* that literature generally is not game, and I think he is right. No one has ever, or will ever, presumably, mistake Lewis Carroll's *Alice* books for elaborate chess problems, let alone chess games, as important as chess can be to certain readings of those novels. Novels are novels, and games are games: they operate differently; they are used differently.[2] On the other hand, I think it is also true that notions of game and gamefulness are important in providing the structures and contexts that are necessary for the postmodern novel—indeed, for the novel itself—to exist.

It is not surprising that the so-called "golden age" of the English-language novel arose in the midst of a highly stable, highly structured, and highly ordered age, that of Victorian England, or that the prominent mode of the fiction of that age should have been realism. However, and to continue to simplify greatly, without the external structures that are provided in contexts like Victorian England, the novel needs to construct its own structure, to erect its own framework: for whatever else the novel may be, it is an elaborately constructed form.

I shall argue that structure is provided in American literary postmodernism by game and gamefulness. On reflection, in fact, it is not surprising that game should provide such a structure: we live in a society that is, after all, obsessed with games, sports, and contests of all kinds.[3] Game, sport, and contest provide meaning in the lives of many people in our own age; it is fitting, therefore, that they should also provide structure and meaning to literary postmodernism itself.

CHAPTER TWO

Postmodernism: Definitional Deliberations

> And further, by these, my son, be admonished: of making many books there is no end, and much study is a weariness of the flesh. (*Ecclesiastes* 12:12)

The sheer number of volumes on postmodernism lends a kind of ponderous weightiness to a debate that promises in all of its self-devouring complexity to be capable of lasting past the end of this millennium and well into the next. It is a debate, indeed, plausibly generated and fueled by that august occurrence (Fredric Jameson claims that "an inverted millenarianism, in which premonitions of the future, catastrophic or redemptive, have been replaced by senses of the end of this or that ... [When] taken together ... perhaps [these] constitute what is increasingly called postmodernism" [*Postmodernism* 1]).

It is true that many discussions of the postmodern are couched in fairly apocalyptic terms (postmodernism is regularly associated with such declarations as the end of the book, the death of the author, the decay or possibly the final disintegration and disappearance altogether of literature).[1] The materials of which literary postmodernism is made can be sentimentally nostalgic on one hand and resolutely avant-garde on the other (see, for instance, in recent American film, *Forrest Gump*[2] and *Natural Born Killers*[3]), and sometimes unapologetically both, as those two films also demonstrate. Discussions of postmodernism continue to multiply, celebrate, or decry endings without closure, change without progress, "pastness" without history. Postmodernism as a term is at once ridiculously backward-looking and forward-peering as it celebrates and deplores its own self-contradictory nature; as Allen Thiher writes, "'Postmodernism' suggests at once living through something apocalyptic and existing with an identity that is forever subordinated to the past" (227).

In its exaggerated forms postmodernism is taken up with mind-bendingly self-reflexive logical contradiction: "Post-Contemporary Interventions" is the title of a series of critical works edited by Stanley Fish and Fredric Jameson and published by Duke University Press. Whether the title of the series is descriptively appropriate or merely meant to appeal to the baser instincts of potential consumers of the series or whether there is any real difference between the two is, I suppose, moot. That the question can and should be asked, without serious hope of definitive answer, stands as yet another paradoxical marker of the postmodern.

Scholarly debates about postmodernism circle around postmodern architecture, postmodern art, postmodern dance, film, history, literature, music, philosophy, photography, politics, theatre, theology; Stanley Fogel has recently published a book entitled *The Postmodern University*; Stanley Trachtenberg has edited a collection entitled *The Postmodern Moment: A Handbook of Contemporary Innovation in the Arts*; Arthur Kroker, Marilouise Kroker, and David Cook have produced a *Panic Encyclopedia: The Definitive Guide to the Postmodern Scene*, in which Kroker, Kroker, and Cook invite "panic readers" to share their "panic flash[es]" (14) for a projected (presumably also "definitive") *Panic Volume II*. Postmodernism, then, is not (just) a period, a term, a movement, etc., but (also) a style of education, a moment, a handbook, an encyclopedia, an invitation to panic (which is, as Kroker, Kroker, and Cook point out, "the key psychological mood of postmodern culture" [13]). Postmodernism, in fact, begins not so much to have definitions and descriptive characteristics, but a personality:[4] postmodernism is close kin to "Trout Fishing in America" in Richard Brautigan's *Trout Fishing in America* (itself, of course, part of the putatively anti-canonical postmodernist canon).

The impressively substantial array called postmodernism continues to become vaster and vaster, as vaster and vaster enterprises both map and model it: Jameson expresses amazement at "the way in which virtually any observation about the present can be mobilized in the very search for the present itself," and muses that "the success story of the word postmodernism demands to be written, no doubt in best-seller format" (*Postmodernism* xii, xiii). Or perhaps it could be a television mini-series.

Indeed, an excellent example of postmodernism's conspicuous, voracious consumption and growth can be seen in Jameson's own narrative of the postmodern. *Postmodernism* began life as a "talk, portions of which were presented as a Whitney Museum Lecture" that transformed itself into a short essay ("Postmodernism and Consumer Society" 111n.); that short essay consumed evidence, gained a more portentous title, tripled in size, and was reborn as a long essay ("Postmodernism, or the Cultural Logic of Late Capitalism"); the long essay, in turn, consuming any and all evidence as it went, next contributed the title and appeared as the leading display in a substantial book, itself a kind of catalogue of postmodern displays ranging from architecture to economics, meanwhile constantly both theorizing and berating the postmodern as "this prodigious rewriting operation, [in which] ... everything is grist for its mill and ... analyses like the one proposed here are easily reabsorbed into the project as a set of usefully unfamiliar transcoding rubrics" (*Postmodernism* xiv). The complete display is impressive indeed, as more and more evidence is accumulated to illustrate postmodernism as itself the illustration, the "cognitive map" of "a strange new landscape" (ibid. xxi); more impressive yet is the note that "the materials assembled in the present volume constitute the third and last section of the penultimate subdivision of a larger project entitled *The Poetics of Social Forms*" (ibid. xxii). The growth of this program, in combination with Jameson's announced general suspicion of the motives and methods of postmodernism, displays in the middle of Jameson's work on the postmodern an instructive gap. Postmodernism is seen as indicative of the last stages of decadent capitalism, as a cheapening and commodification of culture, as a rather vulgar example of consumption run amok; at the same time, Jameson's own project commodifies any and all cultural artifacts in its path, turning everything into evidence, and that evidence into the raw materials that feed its own growing narrative. Jameson's *Postmodernism* is itself, ironically, then, one of the better examples available of postmodernism at work: its relentless rewriting and its constant acquisition of commodified artifacts, betray it as its own leading display, exhibiting *par excellence* the same restless consumption that it decries elsewhere.

Michel Foucault cautions that "It is obvious that the archive of a society, a culture, or a civilization cannot be described exhaustively; or even, no doubt, the archive of a whole period" (*The Archaeology of Knowledge* 130). This study is not, then, an attempt to do what Foucault has already declared to be impossible: I have no ambition to be a definitive cartographer of the entire postmodernist landscape. However, I do wish to sketch certain intersections and gaps between and among various discourses of the (mainly literary) postmodern—discourses that Robert Wilson, borrowing his terminology from Michel Foucault[5] and from discourses on museums, calls the "archives" of postmodernism. And I shall suggest a model that may help to make sense of some of the aspects of the collected archives that we call postmodernism.

The archives of postmodernism, notes Wilson, are constituted by "two separate, and differently organized, baggage-trains" ("SLIP PAGE" 99). One archive is a "period-term, esurient in its consumption of evidence, exiguous in its production of interpretation, to name NOW. All culture reflects the economic forces that have created the conditions of its possibility" (100). The other archive is "a highly flexible analytic-descriptive term capable of isolating conventions, devices and techniques across the range of all the cultural products ... that can be caught in a widely flung transnational net" (102). The two archives, though "each [is] largely blind to opposed discursive formations" (112), do intersect; they "overlap and coincide," forming "a nexus of intersecting discourses" (105, 112).

Wilson claims, in short, that postmodernism is a "paradigm-case of the problem of boundaries and slipping categories" (104). I propose in this study to investigate some of the archives of literary postmodernism and the "nexus of intersecting discourses" that runs between them. I shall argue that perhaps the most important terms in the various discourses that constitute postmodernism are "boundary" and "border," terms that appear frequently in postmodernism as themes, as strategies, as obsessions.[6] My proposal is that postmodernist literature is a literature caught precariously on the various edges between and among radically different discourses which express themselves in and through (potentially) radically different media. This book has as its ambition the end, not of seeking or constructing a taming conjunction or reconciliation or synthesis of the many archives of postmodernism (there are, of course, not only two, but many different possible archives, depending on one's organizing principles), but of pointing to intersections and overlaps and gaps and suggesting a model that can conceivably accommodate multiple archives. Thus this book posits boundary negotiation as an exercise that not only "occurs in" the postmodern but that is itself postmodernism in important senses.

Postmodernism will be seen to inhabit the tenuous and slippery boundaries between authors and readers, texts and contexts, metonymy and metaphor, modernism and whatever comes after that, fiction and non-fiction. That is, postmodernism will appear in many or all of the other discourses that make up the sets of conventions by which we run our lives: economics; philosophy; history; painting; photography; film; television. I shall argue that any satisfactory definition of the postmodern must begin by addressing multiple archives, and that any definition that does not involve more than one archive is not necessarily wrong, but is necessarily incomplete. I shall claim, finally, that postmodernism constitutes itself as a game, though I make no such claim for any other kind of literature. I do not wish to argue that texts in general are games, but that postmodernist literature in particular negotiates the boundaries it constructs and of which it is constructed by constituting itself in gamelike ways. That analogy may not apply to any other sort of literature; if it does not, and my case is convincing, gamefulness may turn out to be an important part of the specific definition of what makes up what is called the postmodern in literature.

First, then, I shall enumerate some of the more prominent theories of the postmodern, divided into two archives, under the assumption that the boundary that has been drawn between the two archives is itself slippery, fuzzy, and mutable, and subject to transgression by either side. This summary is not meant to be exhaustive (with new theories appearing daily, it seems, what *could* be exhaustive, except perhaps an on-line, continuously updated catalogue?), but it is meant to be generally representative.

1. Periodizing Hypotheses

Fredric Jameson, though he lacks Charles Newman's almost evangelistic fervour (see below), agrees with Newman and others that the postmodern is generally taken to signal "the end of this or that," and that "aesthetic production today has become integrated into commodity production generally" (*Postmodernism* 1, 4). Postmodern culture, on Jameson's account, displays initially "a new depthlessness ... a weakening of historicity ... a whole new type of ground tone—what I will call 'intensities'—which can best be grasped by a return to older theories of the sublime ... [and] the deep constitutive relationships of all this to a whole new technology, which is itself a figure for a whole new economic world system" (6). Jameson's argument is "not to be read as stylistic description, as the account of one cultural style or movement among others.... Rather ... [as] a periodizing hypothesis, and that at a moment in which the very conception of historical periodization has come to seem most problematical indeed" (3).[7] What Jameson proposes is more properly called, therefore, a "genealogy" than anything which might be construed as "traditional ... linear history" (3). Thus he produces a genealogy of the postmodern that points to its own a-historicity and its preoccupations with present space, to demonstrate that it appears to be, at least at first, essentially a "mutation": postmodernist architecture, for instance, has, for Jameson, "finally succeeded in transcending the capacities of the individual human body to locate itself, to organize its immediate surroundings perceptually, and cognitively to map its position in a mappable external world" (38, 44). Jameson concedes that this "alarming disjunction" is initially bewildering, but can "itself stand as the symbol and analogon of that even sharper dilemma which is the incapacity of our minds, at least at present, to map the great global multinational and decentred communicational network in which we find ourselves caught as individual subjects" (44). The de-personalizing and de-centring of postmodernist art is thus not merely a "characteristic" of the art (as it was for modernism), but *the* characteristic point, precisely the problem. Postmodernism for Jameson becomes a kind of superficial "image addiction," something that can and should, however, "be read as [a] peculiar new [form] of realism," a disjunctive art responding to a disjunctive reality (49).

Jameson is no fan of postmodernism, though he sees little that could seriously be called "deplorable and reprehensible" in postmodernism, finding "old-fashioned ... indignant moral denunciation ... unavailable" to critics, because they are themselves suffused in the postmodern age (46).[8] Jameson ends the chapter of his book that is devoted to postmodern culture, unexpectedly, on what he obviously considers to be a rather positive note:

> This is not then, clearly, a call for a return to some older kind of machinery, some older and more transparent national space, or some more traditional and reassuring perspectival or mimetic enclave: the new political art [needs to achieve] ... a breakthrough to some as yet unimaginable new mode of representing ... in which we may again begin to grasp our positioning as individual and collective subjects and regain a capacity to act and struggle which is at present neutralized by our spatial as well as our social confusion. (54)

Postmodernism, for Jameson, if confusing at the moment, is confusing precisely because we are all submerged in it; if it seems empty, that is because it is making its way to some other, newer, better form of representation; if it is a challenge, it is an exhilarating one which may lead to new "as yet unimaginable" forms of committed political art. If not infused with Newman's sour pessimism, then, Jameson's discussion is still populated by images of ruptures, breaks, changes, discontinuities, transitions—the same borders and boundaries—edges—that are ubiquitous in Newman's work. Both Newman and Jameson agree, furthermore, that postmodernism commodifies culture, making of culture a capitalist product, to the obvious detriment of culture. However, the "depthlessness ... [and lack of] historicity" (6) and so on that they both detect in postmodernism, for Jameson constitute a stage in a kind of genealogy, a crossroad in a strange landscape, whereas for Newman, they signal an end of "real" culture. For Jameson, there is possible life after postmodernism; for Newman, postmodernism merely signals the final stage in the lingering death of modernism, and of Western culture with it.

Charles Newman's *The Post-Modern Aura* decries postmodernism ("the very term signifies a simultaneous continuity and renunciation" [5]) as symptomatic of latter-day, decadent capitalism, characterized by what Newman calls "*climax inflation*—not only of wealth, but of people, ideas, methods, and expectations—the increasing power and pervasiveness of the communications industry, the reckless growth of the academy, the incessant changing of hands and intrinsic devaluation of all received ideas" (6). "The overwhelming sense not merely of the relativity of ideas, but of the sheer quantity and incoherence of information, a culture of inextricable cross-currents and energies—such is *the* primary sensation of our time.... The lens turned on contemporary art must allow for all distortions" (9). Newman's apocalyptic vision of postmodernism as "this black hole in the present" (36) argues that even "if there is no body of ideas so elegant as to constitute a true *Zeitgeist*, there are ... a number of shared questions and feelings, abortive procedures and mock dead ends" which go together with certain "attitudes, antipathies, historical views and even aesthetic choices" to characterize the postmodern "period" (13, 36). The postmodern period is, he claims, "the first period which does not idealize some specific historical period as an emulative model, or attempt to recapture the purity, however illusory, of some vanished age"; this lack of idealism is, in Newman's view, indicative both of a lack of a sense of history, and a lack of true culture (39). "It is this innocent fact, of a *real* non-existent audience and a *genuine* indifference to Art, which the Post-Modern artist confronts if not always honestly, at least consciously" (50).

Newman borrows John Barth's term and points disparagingly to the "used-upness" of postmodernism, the bankruptcy of current "so-called culture," regretting its shallowness, ruing the "period" even while he discusses it. Newman's argument hinges on at least two important points: first, he is sure that the postmodern indicates an apocalyptic change, a radical break with "the past." He sees that past as superior in seriousness, coherence, and in its sense of and appreciation for both history and "true" art. Second, he links literary decline with economic decline, as evidence pointing here to a publishing industry that is obsessed with books as mere products, and there to writers who have been forced into teaching out of economic necessity ("the proper question is not whether this has affected writers, but whether this is the best way to make use of our writers" [127]). Newman sees postmodernism as an unfocused, unrisky, cliched, decadent, and empty form:

> In short, art has become indexed to cultural inflation, and the only risky thing about art in such a context is to possess an asset which is not subject to irrational speculative bidding. It is no accident, then, that the deterministic cliches of our time are above all *perceptual*, the most impermeable ideology yet devised, in which the artist must always provide "new ways of seeing," and that such "sight" dominates and determines the external world. Taken far enough, this exempts artists from having to produce ideas, values, methods, or judgment. And it is equally handy for the consumer, for it provides him with an all-purpose rejoinder: "it all depends on your point of view"; meaning, of course, that he has no point of view. (180)

This lack of centre, history, stability, tradition, point of view amounts to an intricate ploy to promote consumption: "As Gerald Graff puts it, 'advanced capitalism needs to destroy all vestiges of tradition, all orthodox ideologies, all continuous and stable forms of reality in order to stimulate higher levels of consumption'" (51).

Leaving aside the interesting postmodernist technique that Newman himself uses here—the acknowledged but undocumented quotation from Graff inserted as a playful bit of intertextuality that is tantalizingly out of reach, unstable, unverifiable—Newman's configuration of the postmodernism constructs it as unremittingly empty, shallow, and bleak, and, to its disadvantage, based on image rather than word. Newman's postmodernism is the result of a rupture with the past; it is besieged by indiscriminate relativity, a lack of history, and a paucity of ideas. If his postmodernism plays, it does so in an inane and soulless vacuum.

Terry Eagleton is also, along with Jameson and Newman, best represented in this first archive; in his "Capitalism, Modernism and Postmodernism" Eagleton notes with dismay "the depthless, styleless, dehistoricized, decathected surfaces of postmodern culture," seizing on the logic of the commodification of art as a prime symptom of what is wrong with postmodernism (61).

According to Eagleton, postmodernism "mimes the formal resolution of art and social life attempted by the avant garde, while remorselessly emptying it of its political content"; it does this, however, not in a traditional parodic mode, with ulterior motives and an ironic sense of history, but cynically, "alienating us from our own alienation" by claiming that there is no intelligible "dream of authenticity" (61) from which to be alienated:

> Post modernism takes something from modernism and the avant-garde, and in a sense plays one off against the other. From modernism proper, postmodernism inherits the fragmentary or schizoid self, but eradicates all critical distance from it, countering this with a pokerfaced presentation of "bizarre" experiences which resembles certain avant-garde gestures. From the avant-garde, postmodernism takes the dissolution of art into social life, the rejection of tradition, an opposition to "high" culture as such, but crosses this with the unpolitical impulses of modernism. (72)

According to Eagleton's reading of late capitalist logic, this empty combination of gestures in postmodernism results in "social reality [being] pervasively commodified"; "if the artefact is a commodity, then the commodity can always be an artefact" (62).

Eagleton thus sees postmodernism not as the deeply pessimistic grist for a Newmanesque mill, nor, as does the cautiously optimistic Jameson, as a transitional cause for new hope for a new kind of committed political art, but as a reactionary form of anti-modernism that replaces political commitment with commodification, and genuine artistic statement not exactly with parody, but with caricature. Eagleton concludes his discussion with the claim that "An authentically political art in our time might similarly draw upon both modernism and the avant-garde, but in a different combination from postmodernism" (72–73).

Although his conclusions are different from theirs, Eagleton has in common with both Jameson and Newman the sense that postmodernism really does represent a profound rupture with the past, and that it displays therefore a loss of both history and political commitment. Postmodernism, for all three, is a resolute late-capitalist commodification of art, with all the shallowness and superficiality that is suggested by those terms.

Jean-François Lyotard's theory of postmodernism, too, fits into this archive: for Lyotard the postmodern is an "age," a "condition of knowledge" (3, xxiii). Lyotard argues that this postmodern "condition of knowledge" is splintered and performative, and based on power, commodity, production, performance, and, ultimately, localised rather than generalised language games: "Simplifying to the extreme," writes Lyotard, "I define *postmodern* as incredulity toward metanarratives" (xxiv). Boundaries between and among systems are blurred or erased; definitions become inevitably localised and situational; standards disappear, to be replaced by power systems that shift according to context. "The grand narrative has lost its credibility, regardless of what mode of unification it uses, regardless of whether it is a speculative narrative or a narrative of emancipation" (37). Lyotard's postmodernism stresses performance: the postmodern is a spectacular demonstration of the impossibility of "metanarratives," by which Lyotard means any formulation of absolute standards or rules or categories that stand outside or beyond the immediate performance in question. Lyotard's work, however, criticises the nostalgia that derides postmodernism and favours a return to realism and "realistic" master narratives, or that sees the artist as either representative of a "community" or else its saviour:

> It is not necessarily the same thing to formulate a demand for some referent (and objective reality), for some sense (and credible transcendence), for an addressee (and audience), or an addressor (and subjective expressiveness) or for some communicational consensus (and a general code of exchanges, such as the genre of historical discourse). But in the diverse invitations to suspend artistic experimentation, there is an identical call for order, a desire for unity, for identity, for security, or popularity.... (73)

Any "security" in art, any master narrative, any "rule" of conduct, must, in Lyotard's estimation, be constructed; in the postmodern age, they are not givens: "those rules and categories are what the work of art itself is looking for. The artist and the writer, then, are working without rules in order to formulate the rules of what *will have been done*. Hence the fact that work and text have the characters of an *event* ..." (83). One of the differences, then, between modernism and postmodernism, in Lyotard's estimation, is that the modernist or the realist might break rules that already exist; the postmodernist, on the other hand, is not a rule-breaker, but always, ever, and inevitably, a rule maker, and necessarily so, because of the conspicuous lack of pre-given rules that the now extinct "master narratives" once supplied.

As Wilson wryly notes, "writers within the first archive seldom evince much enthusiasm for postmodernism" ("SLIP PAGE" 101). Jean Baudrillard might be a shiny exception to that provisional rule. Baudrillard apparently takes his position as a postmodernist theorist who is engulfed in postmodernism seriously; rather than try to stand outside the (largely empty and superficial) discourse he theorises, he climbs aboard. In Baudrillard's view, there can and must be no distinction between the postmodern object and its representation, the thing and the idea. He envisions a world of "hyperreality," "simulacra," and "simulations," where images have taken over from linear, reasoned, Platonic discourse to create a new, postmodern, "ecstatic" mode of communication. He writes:

> These would be the successive phases of the image:
> 1 It is the reflection of a basic reality.
> 2 It masks and perverts a basic reality.
> 3 It masks the *absence* of a basic reality.
> 4 It bears no relation to any reality whatever: it is its own pure simulacrum.
>
> In the first case, the image is a *good* appearance: the representation is of the order of sacrament. In the second, it is an *evil* appearance: of the order of malefice. In the third, it *plays at being* an appearance: it is of the order of sorcery. In the fourth, it is no longer in the order of appearance at all, but of simulation.
>
> The transition from signs which dissimulate something to signs which dissimulate that there is nothing, marks the decisive turning point. The first implies a theology of truth and secrecy (to which the notion of ideology still belongs). The second inaugurates an age of simulacra and simulation, in which there is no longer any God to recognize his own, nor any last judgment to separate true from false, the real from its artificial resurrection.... This is how simulation appears in the phase that concerns us: a strategy of the real, neo-real, and hyperreal, whose universal double is a strategy of deterrence. ("Simulacra and Simulations" 170–71)

Baudrillard's formulation is the end result of the disappearance of the word as the final unit of textuality, of the text as final authority, and of the author as final maker of the text. "One way of describing Baudrillard's project is to see it as a species of inverted Platonism.... It is no longer possible to maintain the old economy of truth and representation in a world where 'reality' is entirely constructed through forms of mass-media feedback" (Norris, *What's Wrong* 166). The postmodern world of Baudrillard is filled only and ever with self-referential signs, and Baudrillard celebrates these signs and this tendency, labelling those who do not so readily celebrate as nostalgic and naive: "There is no longer any transcendence or depth, but only the immanent surface of operations unfolding, the smooth and functional surface of communication. In the image of television, the most beautiful prototypical object of this new era, the surrounding universe and our very bodies are becoming monitoring screens" (*Ecstasy* 12). The dismissive scepticism toward postmodernism displayed by most residents of this first archive is entirely absent from Baudrillard; he discusses postmodernism as "a consequence of the disappearance of causes and the almighty power of effects" (*Fatal Strategies* 37) and acknowledges that "today there is no longer transcendence, but the immanent surface of the development of operations, smooth surface, operational, of communication" (ibid. 66). Whereas Jameson, while claiming that postmodernism must be "confronted as a historical situation rather than as something to be morally deplored or simply celebrated" (Stephanson 12), nonetheless yearns for a renewed commitment to politics, values, and morality in art, Baudrillard's discussion eschews such yearnings, instead embracing postmodernism as the art of the present. Baudrillard does not "simply celebrate" the postmodern; though he does celebrate, he also confronts it, expands upon it, plays with it, explores it, refusing the moral indignation entirely which so many of his peers, explicitly or implicitly, espouse, opting instead for the elliptical irony for which he is now so (in)famous:

> *À propos* the cinema and images in general (media images, technological images), I would like to conjure up the perversity of the relation between the image and its referent, the supposed real; the virtual and irreversible confusion of the sphere of images and the sphere of a reality whose nature we are less and less able to grasp. (*The Evil Demon of Images* 13)

As a response to critics who claim that such a confused, endlessly deferred, and superficial discourse is also therefore without meaning, Baudrillard insists that that is not the case; it's just that meaning does not pre-exist its formulation and cannot be separated from that formulation: "everywhere one seeks to produce meaning, to make the world signify, to render it visible. We are not, however, in danger of lacking meaning; quite to the contrary, we are gorged with meaning and it is killing us" (*Ecstasy* 63). Baudrillard's postmodern world, then, is based on image; it is characterised by "simulation" and "simulacra," which do not re-present reality, but have displaced it, and therefore present their own realities; Baudrillard's postmodern world is a vortex of endlessly deferred, ultimately absent final meanings, and endlessly invented provisional ones; it is a world where television, not the book, is not only "beautiful," but the "prototypical" form of discourse.[9]

2. Hypotheses of Technique

"The second archive constructs postmodernism as a highly flexible analytic-descriptive term, capable of isolating conventions, devices and techniques across the range of all the cultural products ... that can be caught in a widely flung transnational net" (Wilson, "SLIP PAGE" 102). This second archive tends to attract postmodernism's enthusiasts and apologists.

Larry McCaffery, for instance, in his *The Metafictional Muse*, suggests that postmodernism is connected intimately with "a well-established tradition of avant-garde art" (x). Postmodernism for McCaffery is subversive of external systems of operation and organization precisely because it is self-consciously, and pointedly, aware of the subjective and ultimately fictional nature of all human systems of organization. In McCaffery's view, only provisional and internal rules of consistency are—or can be—important for, or applicable to, discussions of postmodernism.

If the focus of postmodernism, as McCaffery claims, is inward-looking, it is not surprising that the metafictional aspects of postmodernism are for him its most significant aspects: it is not surprising for McCaffery to discover, for instance, that postmodernists have tended "to focus not on reality but on the imagination's response to reality—a response which [is] the *only* aspect of reality which could ever be known" (13). On one hand, this is indistinguishable from a textbook explanation of modernist stream-of-consciousness techniques, or various facets of Romanticism. On the other hand, it is a discussion that disagrees strongly with theories like

Baudrillard's, continuing, as it does, to separate reality and the artist's response to reality. Whereas for Baudrillard, reality has been displaced, for McCaffery, the response to and interpretation of reality have been skewed and fractured, in ways that have a decidedly modernist (and avant-gardist) slant.

McCaffery tends to conflate "postmodern" with "metafiction" and then to identify metafiction (and therefore postmodernism) through shared themes and techniques: metafiction/postmodernism is "disruptive," "radical," "experimental"; it deals with "the hazards of knowing"; it is liable to "confound fact and fiction," and "flaunt its [own] artifice." It is, in short, "that type of fiction which either directly examines its own construction as it proceeds or which comments or speculates about the forms and language of previous fictions" (16). It is unclear how these formulations can be usefully applied to the work of Coover or Barthelme, and not to *Don Quixote* or *Tristram Shandy*, for instance, or what, in short, even if "examples of this type of fiction are easy to spot" (16), makes postmodernism distinctive in any way. Postmodernism, for McCaffery, appears essentially to be metafiction, but histories of metafiction are entwined intimately, from their beginnings, with histories of literature (and postmodernism is lacking in literary histories, until very recently).

McCaffery posits a postmodernist "shift in literary sensibilities" to explain the sudden appearance of the postmodern, and his explanation for that "spectacular shift in literary sensibilities" has a good deal to do with the "atmosphere of rebellious energy" that so characterized the nineteen-sixties (19). It is largely to the 'sixties that McCaffery traces postmodernism, at least in its American manifestation; and largely, within that decade, to a small group of writers "of obvious genius," accompanied by a cadre of "sympathetic critics" (22). The picture emerges from McCaffery, then, of a rather insular group of Americans "of genius" transforming the work of "significant Europeans" (all of them, interestingly, French: Butor, Sarraute, Robbe-Grillet, Queneau) into what is now called postmodernism.

On one hand, McCaffery is guilty of using very slippery, ill-defined terms: "genius" is "obvious" when we see it (and so does not really need definition); "significance" is equally obvious; "literary sensibilities" are monolithic, except for the "rebellious" representatives of the postmodern "avant garde," who are also obvious and "easy to spot." On the other hand, he seems guilty of exactly the kind of foreshortened sense of history—especially literary history—that so bothers critics like Charles

Newman. If Baudrillard and Lyotard and other representatives of the first archive are correct, and various of our "external" systems of measurement have collapsed, then the kinds of structures that the avant-garde has historically played with and against are no longer self-evidently there to be played with. McCaffery's formulation of postmodernism, as useful as it might be as a catalogue of isolated conventions, suffers from a (possibly typical, and if not distinctively, then characteristically postmodernist) lack of history. Surely a listing of techniques alone cannot be used to define postmodernism as something that has not existed before, when all of the techniques themselves have existed for as long as fiction itself has.

For Brian McHale in *Postmodernist Fiction*, postmodernism is best understood when it is considered in the light of Roman Jakobson's theory of "the dominant": postmodernism for McHale represents a "change of dominant," a shifting of concerns away from those of modernism, broadly construed as "epistemological" (9), to concerns more rightly characterized as "ontological" (10). For McHale, then, postmodernism is less a question of how the alienated individual so typical of modernist narratives knows and survives in a hostile world, than it is a question of how a world—any world—can or does exist at all, or how "being" can be constructed in any given (necessarily provisional) world. This shift is invoked to help to explain and structure various "catalogues of postmodernist features," such as those enumerated by David Lodge, Ihab Hassan, Peter Wollen, or Douwe Fokkema.[10] Whether the differences between modernism and postmodernism can really be schematized as succinctly as McHale does (epistemology vs. ontology), or as expansively as Hassan does (with his table of thirty-five differences), or more expansively still, the real usefulness of McHale's work lies in the fact that it is what Wilson calls "a concise encyclopedia of the second archive" ("SLIP PAGE" 103), examining, as it does, the problematic characteristics of postmodernism from within the structure of the "ontological dominant" it sets up.

One of the more frequent criticisms of McHale's *Postmodernist Fiction* is that it is inclined to see postmodernism as fitting too rigidly into a pattern, as a too-neat explanation of a postmodernist "'fixed essence.'" This critique is apparently the one that prompted McHale's second book on the "problem" of postmodernism, *Constructing Postmodernism*. There, although McHale continues "to invest the term with a certain semantic substance," he "insist[s] on the discursive and constructed character of postmodernism" (1). Thus "*Constructing Postmodernism* proposes multiple, overlapping and intersecting inventories and multiple corpora; not *a* construction of postmodernism, but a plurality of constructions, constructions that, while not necessarily mutually contradictory, are not fully integrated, or perhaps even integrable, either" (3). McHale's technique is to tell some of the stories of the postmodern, constructing, re-constructing, de-constructing postmodernism as a series of stories, essays, excursions. It is not surprising, then, to find, as promised, that "there 'is' no such 'thing' as postmodernism.... if what one has in mind is some kind of identifiable object 'out there' in the world, localizable, bounded by a definite outline, open to inspection, possessing attributes about which we can all agree" (1). By the end of *Constructing Postmodernism*, however, one is struck by several features of McHale's work. First, postmodernism is well enough described and defined by McHale as to be capable of being, if not captured, exactly, then at least fairly represented in a certain kind of canon (one that includes, at least, "the usual suspects": Barth, Barthelme, Coover ...). Second, *Constructing Postmodernism* is, in a very real sense, substantially only a presentation of further evidence for the ontology/epistemology thesis of *Postmodernist Fiction*. Third, although at one level, postmodernism is for McHale a constructivist project, postmodernism has, by the end of the book, been thoroughly enough "constructed" that the label "mainstream postmodernism" does not, any longer, seem out of place (227). Postmodernism as described by McHale begins to look again like it might have a "fixed essence," after all.

Allen Thiher's study focuses on the "language games" and related problems that circle in and about postmodernism, in order to achieve what Thiher calls

> ... a certain exemplarity, not in order to find the "essence" of postmodern fiction but to identify a number of common traits that can be called, in Wittgenstein's phrase, "family resemblances." With a sufficient number of these resemblances in hand one can then sketch out an adequate definition of what we mean today by postmodern fiction. (*Words in Reflection* 7)

Thiher's book begins with discussions of the work of such notable modern language theorists as Wittgenstein, Heidegger, and de Saussure, and claims that postmodernism is "grounded" in and moves out of modernist theories of language. He moves from there into the postmodern language theories of Derrida, and such typical postmodern problems and obsessions as representation (a consideration of the bases of postmodernism's "attack on the metaphysics of representation" [91]), voices ("who speaks and from where" [120]), play ("modern language theories agree in telling the storyteller that, whenever he tells a tale, he has already started to play" [156]), and reference ("how language theory and fictions consider the way language can refer to the real" [188]). Thiher's conclusion is that postmodernism exists as "a series of discriminating traits ... that allows us to see ruptures as well as continuities.... [and as, finally,] a counter in our language games" (227).

Thiher's list of "traits" is not exhaustive, and was not meant to be. Thiher aims, he claims, to be useful, not definitive (after all, with the endless play of language, the endless deferral of meaning in the postmodern linguistic universe, who or what *could* be "definitive"?). Thiher's reasoned, scholarly work establishes certain links between postmodernist fiction and various language theories in compelling fashion, and within those constraints, it is exemplary. However, *Words in Reflection* suffers from the same problem that plagues so many versions of postmodernism within this second archive: there are no narrative techniques or concerns that are unique to postmodernism. Neither questions concerning reality and the complex relationships between reality and fiction; nor various related problems of reference; nor playful self-reflexivity are the exclusive province of postmodernism, as anyone who has read Sterne or Smollet or Fielding or Cervantes or Diderot knows well. Thiher knows too, of course, that none of these obsessions or techniques is unique to postmodernism: that awareness is part of the reason he depends on Wittgenstein's "family resemblances." But in making postmodernism a response to modernist language theories, Thiher nonetheless foreshortens history, and moves into a stance that notes family resemblances among the nuclear family members of postmodernism only, leaving cousin Sterne, for instance, out of the family portrait altogether.[11]

As Wenche Ommundsen (for instance) shows, it is not difficult to construct a complete history of metafiction and its techniques: "The relationship between words and things and the various rules governing the representational game have captured the imagination of writers from classical Greece to the postmodernist era" (*Metafictions?* 33), but to claim that "fictional reflexivity is a postmodern novelty" (*Metafictions?* 102) is merely to evade several of the important questions that are posed by a serious consideration of the history of reflexivity, on one hand, and the vaster (though shorter) history of postmodernism, on the other. Ommundsen's warning that "postmodernism ... is ... a category so complex and disputed that its relationship to metafiction cannot be precisely defined" (*Metafictions?* 14) is a wise one. Metafiction may itself turn out to be a collection of techniques, and reflexivity may be usefully studied as a product of certain reading practices, but those techniques and products have histories.

Reading Thiher, one gets a sense of where certain of the characteristic clusters of themes and techniques of postmodernism might have gained their currency; however, for all his meticulous "grounding" of postmodernism in modernist theories of language, Thiher finally opts for a technique-centred theory at the expense of a more comprehensive historical (archaeological? genealogical?) perspective or context.

Linda Hutcheon, on the other hand, is alive to problems of history in the postmodern, although she does not attempt to construct an actual history of postmodernism itself;[12] instead, Hutcheon largely limits her various discussions of postmodernism to what she calls "historiographic metafiction." For Hutcheon, history is the theme, structure, and organizing principle of "authentic" postmodernism. Historiographic metafiction is "intensely self-reflexive" fiction which "paradoxically ... [lays] claim to historical events and personages," while at the same time exhibiting a "theoretical self-awareness of history and fiction as human constructs," which "is made the grounds for its rethinking and reworking of the forms and contents of the past" (*Poetics* 5). The postmodernism that interests Hutcheon works, she says, "within" the conventions of literature, history, and theory in order to "subvert," though not "banish" or "overthrow" them (the distinctions among these are very fine). Hutcheon claims that, at least in literature, neither defining the essence of postmodernism, nor investing it with a transcendental identity is possible. All (postmodern) definitions must be contextual,[13] and hers appear at first to be resolutely so, nearly to the point of being parochial, or even idiosyncratic: "*for me*," she writes, "postmodernism is a contradictory phenomenon, one that uses and abuses, installs and then subverts, the very concepts it challenges" (*Poetics* 3). To a list that includes definitions of postmodernism by McHale, Jameson, Baudrillard, and others, Hutcheon "[adds] *my own* paradoxical postmodernism of complicity and critique, of reflexivity and historicity, that at once inscribes and subverts the conventions and ideologies of the dominant cultural and social forces of the twentieth-century western world. My model for this definition is always that of postmodern architecture and its response to the ahistorical purism of the Modernism of the International Style" (*Politics* 11); "postmodernism, *as I see it*, is ... paradoxical and problematic, as witnessed perhaps by the continuing debates on its definition. It both sets up and subverts.... It uses and abuses ..." (*The Canadian Postmodern* 2).[14]

In the course of setting up her definitions, Hutcheon removes from her discussion "the extreme non-representational textual play and self-reference" of such phenomena as the French "nouveau nouveau roman" and American "surfiction" as having characteristics and preoccupations not pertinent to her problems or interests (*The Canadian Postmodern* 2). Hutcheon's work taken as a whole, then, becomes a vast comparative catalogue of a certain brand of postmodernism, and is potentially useful and helpful within Hutcheon's rather narrow definitional bounds, but her descriptions turn out to be not particularly provisional or contextual after all. Instead, while reading Hutcheon's work, one constantly runs into exclusions like the one proffered for surfiction and the nouveau nouveau roman. Paradoxically, her definitions become far too narrow and her criteria for postmodernity become at once too accommodating and too limiting to do much more than delimit a description of a certain kind, or branch, or canon, perhaps, of postmodernist literature.

The application of Hutcheon's criteria certainly leads to some extremely counter-intuitive results: not only are the nouveau nouveau roman and surfiction stepped around, but in *A Poetics of Postmodernism*, for instance, Rudy Wiebe (apparently because of the historical preoccupations of *The Temptations of Big Bear* and *The Scorched-Wood People*) is displayed as a more important practitioner of postmodernist fiction than, say, Donald Barthelme (who is not mentioned at all in either *Poetics* or *Politics*). Barth's *The Floating Opera* appears in the extensive bibliography of *Poetics*, and his *LETTERS* is a part of the discussion of both *Poetics* and *Politics*, but neither *Lost in the Fun-house* nor *Chimera*, texts often considered as Barth's (Barth himself is often, of course, considered one of the more exemplary of practitioners of postmodernism) most exemplary acts of postmodernism, is mentioned in either of Hutcheon's books. Fowles's *The French Lieutenant's Woman* figures in Hutcheon's discussion, but neither *Mantissa* nor *The Magus* appears. Neither Coover's *Pricksongs & Descants* nor *The Universal Baseball Association, Inc.* is discussed either, although his *The Public Burning*, with its claimed historical basis and pointed examination of certain notions of historical "truth," is a strong force throughout Hutcheon's work.

It is worth mentioning that several of the more conspicuously absent texts I have cited here do appear in Hutcheon's *Narcissistic Narrative*, before her definitions stopped being provisional and hardened into the form they now have, before Hutcheon stopped discussing (merely) "metafiction" itself and turned to "historiographic metafiction" as a synonym for postmodernism. It is, of course, unfair to draw strong conclusions based on the appearance or lack of appearance of particular texts in Hutcheon's bibliography and discussion, but I think it is fair to note that Hutcheon's discussion is framed very strongly by the limits of the model that she sets up: Hutcheon's formulations point to particular characteristics in particular texts and that evidence is used, not exactly to indicate the presence of postmodernism in those particular texts, but to delimit the definitions and sketch the borders of literary postmodernism at large. For Hutcheon, "historiographic metafiction" is not (only) the trace of the postmodern that runs through certain novels, but is itself essentially postmodernism in literature; fictions that are not "historiographic" enough are either ignored or dismissed as not authentically postmodernist (see *Lost in the Fun-house*, on one hand, and surfiction, on the other); so, too, those fictions that are presumably not explicitly "metafictional" enough (see, for instance, anything written by John Hawkes).

The discussions of postmodernism belonging to this second archive—that is, the ones that can be seen to circle around technique—are very often related to clusters of techniques, although that is not always so. For instance, Alan Wilde, whose *Horizons of Assent* belongs in this archive, considers irony to be the informing feature of postmodernism. In order to discriminate between the irony which is so much a part of so many discussions of modernism, and postmodern irony, Wilde posits "[*postmodern*] irony as a mode of consciousness, a perceptual response to a world without unity or cohesion" (2). Whereas for modernism, in Wilde's formulation, irony was a formal characteristic of literature, the final resolution of which was the resolution of (innate) textual conflict, in postmodernism irony becomes the act of writing itself, the creation of order (however provisional and unstable) out of the absolute chaos which is postmodern reality. For Wilde, the provisional and shifting order which writing can bring, even momentarily, to a provisional and shifting world is ironic, and is itself both postmodernist and postmodernism.

Benzi Zhang, in his dissertation on the image of the Chinese box in postmodernist literature, construes self-reflexivity as a very strong indicator of postmodernism; Zhang's formulation incorporates a very pronounced Lyotardian distrust of "master" (or even "stable") narratives as an essential part of self-reflexivity, without conflating or simplifying metafiction and the postmodern to create a stable model. Zhang writes that the reflexive strategies employed in his version of postmodernism "[uncover] a great deal about ... postmodern narrative practice—for instance, the codes by which we organize reality, the means by which we organize words about it into narrative, the ways in which readers are drawn into narrative, and the nature of our relation to the linguistic medium and to the reality behind/inside it" (26). Zhang's codes of the postmodern are not, however, confined (as they are for Hutcheon, for instance) to history and fiction; he is able to discuss not only history and fiction in this context, but also other reality-organizing narratives—stories generally that humans tell each other. In addition, rather than the Chinese box's being exclusively or even predominantly a textual phenomenon, the reflexivity of postmodernism arises for Zhang because "the reader is required to perform successive acts of resignification and recontextualization of previously established meanings" (4). Zhang's formulation moves away from a strictly text-centred critique and toward a reading- and reader-centred activity. Zhang's postmodernism is not, then, a product so much as a process: a process that crosses from one archive into the next, dancing on the edges of discourses, negotiating the nexus between archives.

Wenche Ommundsen, too, has addressed the problems of reflexivity and metafiction in compelling ways. Rejecting, on one hand, the notion that "metafiction is a particular kind of writing" and on the other, that "*all* fiction ... is fundamentally and predominantly self-reflexive," Ommundsen suggests instead that "all fiction carries within itself the potential for a metafictional reading, a latent self-consciousness which can be activated by the reader" ("The Reader in Contemporary Metafiction" 171–72). As Ommundsen notes, metafictional techniques do not "exactly *destroy* the fictional illusion (after all, we knew all along the story was a fiction); [they do,] however, force us to reflect on the nature of that illusion, and on our own complicity in its creation" (*Metafictions?* 9). I have already cited Ommundsen's reluctance to expand her discussion of metafiction into a discussion of postmodernism, but her location of metafiction as contained in the role of the reader is certainly a recurring figure in many discussions of the postmodern.

Indeed, the role of the reader in postmodernism generally is something that would be difficult to over-state. Ross Chambers notes, for instance, that "irony is a function of reading," and writes of "the conditions of indeterminacy we call reading" ("'Narrative' and 'Textual' Functions" 31,

28); Stanley Fish discusses "communities of readers" and their vital role in the construction of meaning in texts (see *Is There a Text in this Class?* and *Doing What Comes Naturally*). The reader in one guise or other is an element in most or perhaps all of the formulations of postmodernism that I have cited thus far; even in formulations from the first archive, the reader as consumer of the products of "late capitalism" is self-evidently attractive; an active reader is vital for equally obvious reasons in Brian McHale's constructivist postmodernism, and in Baudrillard's glimmering authority-free zones of simulacra, and in Thiher's language games, and in Wilde's irony, and in the myriad other moods, attitudes, and assumptions that together constitute the postmodern.

Further, if, as McHale and others have claimed, television is the pre-eminent vehicle of the postmodern, it is itself spectacular vindication of the view that posits the reader/viewer as the final, ironic, and isolated maker of meaning: see, for instance, Postman or Miller or McKibben on the ironic disposition of the postmodern reader/viewer, and the uses to which that irony is routinely put.[15] A concentration on the reader and readers' activities signals a movement of critical interest away from formulations that posit postmodernism either as an age generally, as in Jameson, or as strictly a text-centred phenomenon, as in Hutcheon. A concentration on the reader as well as the text begins to pull the two archives toward each other, and to begin to construct the intersection that wavers between the two.

This book takes as its starting point, then, the view that postmodernism is at best a functional term, and at worst a meaningless functional term; it is both a period description *and* a set of traits and techniques; it is a protean rhetorical construct, capable of assuming virtually any shape, inhabiting various fundamentally antagonistic camps, and capable of being discussed in each. I think postmodernism encodes into itself deliberately and explicitly the recognition that reading is not a "spectator sport," but an active, participatory one. The various webs of significance and insignificance that together constitute postmodernist intertextuality, for example, are as important to postmodernism as they seem to be problematical for postmodernist theorizers.

Thus an apparent lack of direction in a particular text, the surfeit of uncertainties that that text embraces, the number of references that seem to lead nowhere, the equal number of references that turn out suddenly to seem essential, the explicit acknowledgement that postmodernism is always both context and construct, are all important in modelling and mapping the difficult terrain that is postmodernist literature. A model of postmodernism as game would allow one to read almost any novel as postmodernism; to see any reading experience as active, rather than passive; and to see any text as speculative, tentative, unfinished, unclosed, and not only subject to interpretation, but dependent on it. (See, in this last context, Readings and Schaber, *Postmodernism Across the Ages*.)

Before I leave this part of this study, I should like to make one final observation about the two archives that I have represented here. Those archives are organized as they are not because that is the only possible organization, but because it is one of the more convenient of many possible sets of fictions, or one of many games, that can make sense of them. Other ways of organizing the same theorists and theories include placing all philosophers and cultural theorists—those who tend to cross disciplinary boundaries to track postmodernism at large—in one camp, and those who tend to see the postmodern from within a particular discipline (say, literary or architectural theory) in another.

For instance, Jameson's aim is to describe (his term is to map) the cultural landscape of postmodernist America. Although he draws examples from various disciplines, Jameson is not primarily interested in those disciplines themselves. That is, though he is interested in economic models as they lend credence and heft to his argument, Jameson is not primarily an economist, and is not interested, particularly, in defining and delimiting the field of economics, even of postmodern economics;[16] Jameson is a cultural critic and theorist. He uses film, say, or discussions of literature as examples, but his concern is with constructing an argument about and defining postmodernist culture as a whole. Hutcheon, on the other hand, is a literary theorist. She draws inspiration and theoretical frameworks from architecture (in particular from the work of Charles Jencks), for instance, but she is interested specifically in the literary postmodern: her work on postmodernism is always tied to, and returns to, literature and literary concerns.

Alternatively, Margaret Rose divides theorists of the postmodern into "positive" and "negative" camps (*Parody: Ancient, Modern, and Post-Modern* 198f.); or into groups of those who investigate postmodernism in terms of "double coding" and those who are "deconstructionist" in their pursuits (ibid. 195). Hal Foster sees both "a postmodernism which seeks to deconstruct modernism and resist the status quo and a postmodernism which repudiates the former to celebrate the latter: a postmodernism of resistance and a postmodernism of reaction" ("Postmodernism: A Preface" xi–xii).

I think it is equally possible to divide the world of postmodernist theories into those which celebrate play and playfulness and the shifting, mutable worlds that those terms imply (see Baudrillard, Edwards, et al.); and those which trace the borders, the boundaries of what is and is not properly "postmodern" (see Hutcheon, Jencks, et al.). The various archives taken together will begin to form what can be seen as the postmodern: an attitude of play, performance, and process, combined with borders, boundaries, and rules (however shifting and mutable). Put together, the archives suggest game and gamefulness; that is where I shall turn next.

CHAPTER THREE

Play, Game, Rule, Convention, and Other Ludic Oddities

> To be entangled in hundreds of pages of a writing simultaneously insistent and elliptical, imprinting, as you saw, even its erasures, carrying off each concept into an innumerable chain of differences, surrounding or confusing itself with so many precautions, references, notes, citations, collages, supplements—this "meaning-to-say-nothing" is not, you will agree, the most assured of exercises. (Derrida, *Positions* 14)

1. Play and Literature

Eugen Fink writes that "play is an essential element of man's ontological makeup, a basic existential phenomenon" (19). While Fink may be right in calling play "essential," it is also true that until quite recently, play has been seen with suspicion, and its theorists viewed as triflers who were lacking in seriousness (indeed, that opinion still surfaces from time to time[1]).

Ruth Burke claims that "the history of play theories in the narrow sense begins with ... Friedrich Schiller" (Burke 1). And Mihai Spariosu essentially agrees, writing that the "most important result [of Schiller's project] ... is the firm reestablishment of play as a legitimate topic of serious philosophical discourse, thereby bringing about a revolution in the modern history of ... play concept(s)" (*Dionysus* 53–54).

And it is true: Schiller is eloquent on the importance of play and its central role in aesthetics, in particular. He names play, in fact, as one of three fundamental human "drives,"[2] connecting play with reason and beauty as essential elements of aesthetics, and therefore always part of what makes us human. He writes, in the widely quoted "Fifteenth Letter": "how can we speak of *mere* play, when we know that it is precisely play and play alone, which of all man's states and conditions is the one which makes him whole" (105, emphasis mine); further, "man only plays when he is in the fullest sense of the word a human being, and he is only fully a human being when he plays" (107). In short, writes Spariosu, "Schiller's return to play as a useful philosophical concept occurs in relation to Plato, who, after purging it of irrationality and violence, acknowledges it as the 'noblest' activity of Reason" (*Dionysus* 54).

While Schiller may have been the first modern theorist of play, it is important to note the links between his theories and Plato's. After all, Plato was, according to Spariosu, "the first play theorist in Western culture" (*God of Many Names* 169). According to Spariosu, play, for Plato, is always a matter for serious discussion: "Plato is no doubt a serious man, but he needs play in order to remain serious" (ibid. 192); and play is tied to numerous other branches of thought. For instance, Plato links "art and aesthetics with ... non-violent, rational play concepts" (*Dionysus* 30); however, play also has another side, one that Spariosu calls "pre-rational." Plato is aware of an "archaic, ecstatic" side to poetic play (*God of Many Names* 197), and buries it under a blanket of reason.

But while Plato may have been suspicious of play, using it in the service of rationality, it was Aristotle who drove a wedge between play and seriousness. It is Aristotle who argues that the irrational side of poetic play "must be *tamed* before it can be turned into a useful philosophical tool" (ibid., emphasis mine). Aristotle, then, becomes "perhaps the first Western thinker who seeks to separate [play] from and oppose it to serious philosophical activity" (ibid. 234). This Aristotelian split remained largely in effect (and, as I have already noted, still surfaces from time to time) at least until Schiller.[3]

It is Schiller who, according to Spariosu's account, "manages to [re-link] art and aesthetics with the nonviolent, rational play concepts in their Platonic version" (*Dionysus* 30). In other words, Schiller is not only the first modern play theorist, but one who is of pivotal importance. Because of his *On the Aesthetic Education of Man*, published in the late eighteenth century, Schiller is seen by Spariosu, Burke, and many others as "rescuing" play as a legitimate subject for serious philosophical discourse, not only by reaching back to the Greeks and bridging a gap between Plato and Aristotle, but also by reaching forward toward the twentieth century. Not even Schiller, however, for all his determination to reinstall play as a legitimate, even necessary, form, and a reasonable subject of serious discourse, could have foreseen the impact play and discussions of play would have in the twentieth century.[4]

Karl Groos's *The Play of Man* is among the earliest twentieth-century attempts to classify and systematize play as a cultural activity worthy of study. Groos builds on Schiller's theories specifically in developing play as not only a serious topic, but as essential to any serious discussion of aesthetics. In fact, Groos's book raises many of the issues that have become central in later discussions of play and its role in the lives of humans.

Groos divides his theory of play into six separate "standpoints," in this way:

i. Physiological. Groos notes with approval the important physiological role that play plays, in the healthy release of what he calls "superabundant vigour" (361). Play is important physiologically, claims Groos, because it affords the playing individual "relaxation" (361). It is important, furthermore, because what takes place during play is the "discharge of surplus energy and recreation for exhausted powers" (368). Play is important physiologically, finally, because of its role in supporting and recharging "the physical and mental powers of the individual.... It is, in short, preparatory to the tasks of life" (361).

ii. Biological. Groos notes that "the impulses to experimentation, to fighting, chasing, hiding ... and other play[ing]" (370) fulfils an essential biological role in the life of humans. We share a biological need for play with the other animals: "We find in all creatures a number of innate capacities which are essential for the preservation of species" (374). In short, "play is the agency employed to develop crude powers and prepare them for life's uses.... Play depends, then, first of all on the elaboration of immature capacities to full equality with perfected instinct, and secondly on the evolution of hereditary qualities to a degree far transcending this, to a state of adaptability and versatility surpassing the most perfect instinct" (375).

iii. Psychological. According to Groos, the psychological importance of play is tied closely to the "feeling of freedom" that the player enjoys while playing (388). Groos notes that "the more earnest is a man's life, the more will he enjoy the refuge afforded by play when he can engage in sham occupations chosen at will, and unencumbered by serious aims. There he is released from the bondage of his work and from all the anxieties of life" (389).

iv. Aesthetic. Groos notes the very "close connection between play and aesthetics," calling specifically on Schiller's discussions of play and aesthetics to support the notion that the aesthetic importance of play is "rooted in playful experimentation and imitation" (394).

v. Sociological. For Groos, play serves an important sociological function, at least in part because play leads to an "increased sense of fellowship" among players of games (357). Because humans are not governed as firmly by instincts as other animals seem to be, socializing mechanisms are central to our existence; play is among the more important of those mechanisms.

vi. Pedagogical. "Instruction may take the form of playful activity, or, on the other hand, play may be converted into systematic teaching. Both methods are natural to us" (399). Groos continues: "while disapproving totally of all trifling in education, we still maintain that the school which is conducted exclusively by an appeal to the stringent sense of duty, with no incentive to the higher form of work in which the deepest earnestness has much of the freedom of play—that such a school does not perfectly fulfil its task" (401).

There is, however, a potential paradox in Groos's work: while he acknowledges his debt to Schiller and concedes that "even the most serious work may include a certain playfulness" (400), he nonetheless maintains a fundamental split between play and seriousness. Play is a legitimate topic for serious discussion, but that does not necessarily make play itself serious. What marks play *as* play in Groos's work, in fact, is the boundary between the playful and the serious. Indeed, play is defined by Groos as essentially "activity that is without serious intent" (2); furthermore, "when an act is performed solely because of the pleasure it affords, there is play" (5).

It is worth noting that while Groos maintains that play is characterized by an intrinsic lack of seriousness, each of his classifications of play stresses precisely the *uses* and the *usefulness* of play: play is the not-serious, perhaps, but for Groos it *is* always, nonetheless, (seriously) utilitarian. Spariosu points out a dependence on rationality and utility as part of Groos's fundamentally Aristotelian stance, where "play [is] in the service of rationality ... restrain[ing] its violent, prerational side" (*Dionysus* 176). This tendency once again to subordinate play to seriousness is reflected in the six "standpoints" of Groos's theory, each of which stresses a separate *use* for play.

For Groos, play is virtually ubiquitous in human cultural phenomena; and it is our culture (always tamed, shaped, and controlled by rationality), after all, that civilizes us, that makes us human. Finally, while not all culture is play (for instance, "work" and "duty" and similar culture-forming terms arise throughout Groos's discussion, especially in the section on pedagogical uses for play), all *human* play, finally, is cultural.[5]

Culture is also an important consideration for Johann Huizinga's discussion of play in his very significant book *Homo Ludens*.[6] Huizinga writes that "play and culture are actually interwoven with one another" (5), and that, even though "play is older than culture ... culture, however inadequately defined, always presupposes human society, and animals have not waited for man to teach them their playing" (1). In Huizinga's work, although not all play is culture, all culture presupposes some form of play: "culture arises in the form of play ... it is played from the very beginning" (46).

Although Groos and Huizinga agree about the importance of considering play and culture together, they differ on certain fundamental points. First, whereas Groos wishes to discuss play in its widest uses and in its most general terms, Huizinga is interested specifically in "the relation of play to culture" (7): "We shall not look for the natural impulses and habits conditioning play in general, but shall consider play in its manifold concrete forms as itself a social construction" (4). Huizinga's relatively narrow focus means that "we need not enter into all the possible forms of play but can restrict ourselves to its social manifestations" (7).

Second, unlike Groos, Huizinga does *not* believe that

> Play is the direct opposite of seriousness.... At first sight this opposition seems as irreducible to other categories as the play-concept itself. Examined more closely, however, the contrast between play and seriousness proves to be neither conclusive nor fixed. We can say: play is non-seriousness. But apart from the fact that this proposition tells us nothing about the positive qualities of play, it is extraordinarily easy to refute.... for some play can be very serious indeed. (5)

In fact, Groos's insistence on the utilitarian nature of play amounts to an unacknowledged refutation of his own position: an unwavering utilitarianism, after all, strongly implies "seriousness," or, in Groos's own terms, a lack of play. Huizinga's project, on the other hand, is not an attempt to assign uses to play (though play can "be very serious indeed"), but to describe "the main characteristics of play" itself (7). Huizinga detects four "main characteristics" which, he believes, human play demonstrates:

> i. First and foremost, then, all play is a voluntary activity. Play to order is no longer play: it could at best be but a forcible imitation of it. By this quality of freedom alone, play marks itself off from the course of the natural process. (7)

> ii. A second characteristic is closely connected with this, namely, that play is not "ordinary" or "real" life. It is rather a stepping out of "real" life into a temporary sphere of activity with a disposition all of its own. Every child knows perfectly well that he is "only pretending," or that it was "only for fun." (8)

> iii. Play is distinct from "ordinary" life both as to locality and duration. This is the third main characteristic of play: its secludedness, its limitedness. It is "played out" within certain limits of time and place. It contains its own course and meaning. (9)

> iv. Play ... creates order, is order.... All play has its rules. They determine what "holds" in the temporary world circumscribed by play. The rules of a game are absolutely binding and allow no doubt.... Indeed, as soon as the rules are transgressed the whole play-world collapses. The game is over.... The player who trespasses against the rules or ignores them is a "spoil-sport".... This is because the spoil-sport shatters the play-world itself.... He robs play of its *illusion*—a pregnant word which means literally "in-play" (from *inlusio*, *illudere* or *inludere*). (10–11)

And Huizinga summarizes his theories in this way:

> ... play is a voluntary activity executed within certain fixed limits of time and place according to rules freely accepted but absolutely binding, having its aim in itself and accompanied by a feeling of tension, joy and the consciousness that it is "different" from "ordinary life." (28)

Ironically, Huizinga does precisely what Spariosu claims that Aristotle and many others insisted on doing before they would admit that play could be taken seriously: he calms it, tames it, fences it off with rules and regulations, controls it by setting it apart as a "separate" activity. He makes it, in short, as domesticated and un-play-full as he can. What Huizinga has done, in fact, is to confuse game, with its rules and precision, with play, by making play into an activity that is strongly connected with rules and rule-making (see his point iv), rather than with the anarchic, possibly violent, free, and freeing activity that more recent theorists have insisted that play is and must be.

If Huizinga is guilty of fencing off and confining play too rigidly, he should also be credited, on the other hand, with linking play and imagination, for "acts of the imagination" (136)—that is, technically, the conversion of reality into images (4)—are themselves always, by definition, playful in Huizinga. There have been many theorists who have followed Huizinga's lead in linking play and imagination.[7]

Roger Caillois, in his *Man, Play, and Games*, disagrees with parts of Huizinga's thesis. For instance, Caillois discounts Huizinga's insistence on rules as too rigid: "There is no doubt that play must be defined as a *free and voluntary* activity, a source of joy and amusement" (6, emphasis mine). Whereas Huizinga is preoccupied with confining and delimiting play, with situating it behind and beneath rules, Caillois appears, at least at first, to be more free of such rigidity: "Many games do *not* imply rules. No fixed or rigid rules exist for playing with dolls, for playing soldiers, cops and robbers, horses, locomotives, and airplanes" (8, emphasis mine). These "games," or play situations (Caillois does not clearly separate the two), all conform to the strict injunction that they be both "free" and "uncertain" enterprises (7). Caillois itemizes what he sees as the "formal" qualities of play in this way:

> Play ... [is] an activity which is essentially:
>
> i. Free: in which playing is not obligatory; if it were, it would at once lose its attractive and joyous quality as diversion;
>
> ii. Separate: circumscribed within limits of space and time, defined and fixed in advance;
>
> iii. Uncertain: the course of which cannot be determined, nor the result attained beforehand, and some latitude for innovations being left to the player's initiative;
>
> iv. Unproductive: creating neither goods, nor wealth, nor new elements of any kind; and, except for the exchange of property among the players, ending in a situation identical to that prevailing at the beginning of the game;

> v. Governed by rules: under conventions that suspend ordinary laws, and for the moment establish new legislation, which alone counts;
> vi. Make-believe: accompanied by a special awareness of a second reality or of a free unreality, as against real life. (9–10)

I should note specifically that Caillois includes rules as one of his criteria (see item v). There is, however, a difference between this work and Huizinga's. Where Caillois sees Huizinga as insisting on the "orderliness" of play, Caillois sees rules in his definition as clearing the way for a lack of orderliness, as essentially suspending "ordinary laws." Thus, although he can write in one place that "many games do not imply rules" (8), his fifth point names rules as essential for play: but the rules of play, for Caillois, "suspend ordinary laws"; they do not rigidly legislate behaviour.

While these six are the formal characteristics of play as he sees them, Caillois is not content only to define play in this way; he also constructs a model, a set of scales, to classify games, depending on the kind of play under consideration, on one hand, and on the way a given game is played, on the other. Thus Caillois "propos[es] four main rubrics, depending upon whether, in the games under consideration, the role of competition, chance, simulation, or vertigo is dominant. I call these agon, alea, mimicry, and ilinx, respectively" (12). In addition to these "rubrics," Caillois places games on a continuum "between two opposite poles." This continuum ranges from "paidia" (characterised by "diversion, turbulence, free improvisation, and carefree gaiety") to "ludus" (characterised by "effort, patience, skill, or ingenuity") (12).

Caillois intends his model with its sliding scales and complex rubrics to stand, at least in part, as a critique of Huizinga's theories, and as a kind of map on which a wide variety of human activity can be traced. And as a map and a model of play and game-playing, it is extremely useful: Caillois has managed to construct a useful way to classify and group play and game.

Hans-Georg Gadamer, in his writings on aesthetics (especially in "The Ontology of the Work of Art and Its Hermeneutic Significance," a chapter in *Truth and Method*), wants to situate play as an essential feature of all art. For Gadamer, play is one of the constitutive elements of art. He writes that, although "it is just as true that literature—say in its proper art form, the novel—has its original existence in being read, as that the epic has its in being declaimed by the rhapsodist or the picture in being looked at by the spectator" (160), that does not make the playful element in art merely dependent on phenomenology or on a vague subjectivity. Gadamer is not a naive realist or objectivist; however, he equally does not see play as the territory of reading or writing individuals, but as an integral part of aesthetics, of art, of language itself. Again, he writes:

> I select an idea that has played a major role in aesthetics: the concept of play. I wish to free this concept of the subjective meaning that it has in Kant and Schiller and that dominates the whole of modern aesthetics and philosophy of man. When we speak of play in reference to the experience of art, this means neither the orientation nor even the state of mind of the creator or of those enjoying the work of art, nor the freedom of a subjectivity engaged in play, but the mode of being of the work of art itself. (101)

For Gadamer, play is not something added on, something provided from outside art by a reader or by a context. Play in (say) literary art is not what readers do, or what writers do (although both readers and writers do, in fact, play); play is something that is a fundamental characteristic of art. For Gadamer, in fact, "the work of art is play" (122).

Jacques Ehrmann begins his discussion of play with a critique of Caillois's methodology. He writes:

> Caillois criticizes Huizinga's conception and definition of play as being simultaneously too broad and too narrow. Too narrow insofar as Huizinga retains only one characteristic of play, its competitive aspect.... Too broad insofar as Huizinga fails to delineate with precision the sphere of play, to draw the line between that which, in each culture, belongs to the domain of play and that which belongs to the domain of the "sacred," the "institutional." (31)

Indeed, it is precisely the lines, the borders, the boundaries, and the rules that mark off play from "real life" that ground Ehrmann's critique of Caillois's theory as itself "too narrow," for while Caillois can claim in one place that some play has no (set) rules, he claims elsewhere that play must be bounded by some sort of rule, some sort of law that obtains while the play is playing. Caillois argues against set rules, but is not willing to allow play the freedom to be without control altogether.

Ehrmann, however, wants to construct a model of play that removes it from considerations that mark it off from "real life." He portrays play, not as a separate mode, not as something we do when we're not doing something serious, but as infusing all aspects of human existence. Ehrmann wants to draw a more distinct line between play (which he really does want to discuss) and game (that which is necessarily bounded by rules and limits of space or time, and in which he is not particularly interested) than does Caillois; and he wants to erase the lines between play and reality altogether. For Ehrmann, play as play is beyond rule; it is distinct from game. This distinction—the one between play and game—seems lost on Caillois; it is a distinction, however, that makes Ehrmann's theories possible.

Ehrmann culminates his article "Homo Ludens Revisited" with eight "conclusions." I shall summarize the ones that are important to this discussion:

> ... play is not played against a background of a fixed, stable, reality which would serve as its standard. All reality is caught up in the play of the concepts which designate it.... Each text contains in itself its own reality, which in essence (or by nature!) is put into play by the words which make it up.... Play and reality [are] inseparable.... The distinguishing characteristic of reality is that it is played. Play, reality, culture are synonymous and interchangeable. Nature does not exist prior to culture.... just as culture is, in the last analysis, communication, so is play ... and game. Thus, any theory of communication ... implies a theory of play ... and a game theory.... The player, like the speaker—that is, each one of us—is at once the subject and the object of the play.... Play is articulation, opening and closing of and through language. (56)

Thus Ehrmann wishes to constitute play as essential and ubiquitous, not only as a part of aesthetics, but as a part of human life; indeed, as a fundamental part of reality. With the theories of Ehrmann and others in the late twentieth century, the gap opened by Aristotle begins to close: play slowly begins to be viewed, not with suspicion, but with celebration instead.

And Ehrmann is not alone in his call for a wider application of the play principle to life and art. The theories of Jacques Derrida, for instance, occupy those places where play reaches its anarchic, playful peak.[8] For Derrida, as for Gadamer, and as for Ehrmann, play is a fundamental part of literature, of language; but further, it is a part of meaning itself.

For Derrida, "to risk meaning nothing is to start to play" (*Positions* 14). Meaning arises in Derrida's writing, not in presence, not in being, but in absence, in "a system of differences" (*Writing and Difference* 280), in the "free play" of differences, of the continuing absence of final "theological" and metaphysical meanings, that are always endlessly deferred (*Positions* 14):

> Besides the tension between play and history, there is also the tension between play and presence. Play is the disruption of presence. The presence of an element is always a signifying and substitutive reference inscribed in a system of differences and the movement of a chain. Play is always play of absence and presence, but if it is to be thought radically, play must be conceived of before the alternative of presence or absence. Being must be conceived as presence or absence on the basis of the possibility and not the other way around....
>
> There are thus two interpretations of interpretation, of structure, of sign, of play. The one seeks to decipher, dreams of deciphering a truth or an origin which escapes play and the order of the sign, and which lives the necessity of interpretation as an exile. The other, which is no longer turned toward the origin, affirms play and tries to pass beyond man and humanism, the name of man being the name of that being who, throughout the history of metaphysics or of ontotheology—in other words, throughout his entire history—has dreamed of full presence, the reassuring foundation, the origin and the end of play. (*Writing and Difference* 292)

Derrida is, of course, often attacked as a trifler, as not-serious (the resurrection of that old play/seriousness dichotomy, that criticism of the playful is, in part, what prompted the extended dis-play which makes up *Limited Inc.*), as doing damage to the enterprises of philosophy and of science and of the humanities—indeed, of humanity itself. But what Derrida has done in fact is little more or less than to take the principles of play and playfulness, and apply them to signification, language, discourse, stretching them out as far as they will go, or at least letting them stretch themselves out as far as they will. Derrida (or at least, Derrida's writing) is endlessly playful, turning the theories of Nietzsche, Hegel, Heidegger, Husserl, de Saussure, and Western metaphysics in general into circling language games that are wondrously playful mixtures of philosophy, criticism, and literature. Derrida's writings lucidly illustrate their own point, bobbing, weaving, circling, glorying in demonstrating the play of language, not merely in discussing it. For Derrida, play is a continuing series of endlessly deferred meanings, of dead ends come alive, of illusions, of allusions, of slippery, shifting differences.

This section of this book has not been, and has not aspired to be, an exhaustive treatise on play in literature: many excellent discussions of that topic already exist. Nor has it aspired to be a complete history of play concepts in literature: the beginnings of that history can be found in the work of Spariosu, among others. However, what I have wanted to recognise in this section is, first, the historical alliance between play and literature (and aesthetics more generally): the concepts of play, literature, and aesthetics are linked, and have been so, explicitly, for a very long time. Second, I have wanted to acknowledge and highlight a side of play that has also long been recognized, and often been suppressed. That side is its anarchy; its refusal of neat closure; its evasion of precise definition.

It is this anarchic, open, evasive—playful—side of play that has attracted so much attention recently among literary theorists and philosophers. Indeed, few topics have attracted as much recent attention. And it is true: this is a side of play that is essential to an understanding of postmodernist literature and aesthetics; it has, in fact, been deemed by some to be the essence of postmodernism.

And I believe that play is an essential quality of postmodernism, as wetness is of water. Play needs to be taken seriously in literary studies generally, and in postmodernist literary studies in particular. However, there are other facets to postmodernist literature beside the playful one; it is to one of those facets that I wish now to turn.

2. Game and Literature

Whereas "play" is a topic that has received a vast amount of philosophical and literary critical and theoretical interest, "game" (at least in ways that do not merely use the word as a synonym for play) has been relatively neglected. I should note at the outset that this section of this book is not a discussion of formal game theory, as useful, elegant, or even entertaining as that theory is in certain circumstances.[9] It is not a revisitation, either, of the blurring of distinctions between play and game that I have already noted in the theories of Huizinga and Caillois, neither of whom distinguishes adequately between play and game; to the contrary, the theory I wish to argue for depends upon the play/game distinction.

Brian Edwards has this to say about the differences between play and game in literary analysis:

> Those characteristics which have made "game" fruitful in economics, political science, and resource and military planning—its precise attention to rules of procedure, clear choices and unambiguous consequences—are those which make it inappropriate in literary theory.... Its limitations are apparent: whereas game theory emphasizes rules which determine the manner of play and game, literature is more playful in its ability to transcend, by mockery, parody, or experiment, its own "rules" or conventions; whereas game theory specifies clear choice and unambiguous consequences, literature is unavoidably ambiguous by virtue of its allusiveness and its limited control over reader activity. (15–16)

And he concludes, with regard to game and play and literature: "As the more expansive term, one that subsumes and transcends game, play provides insights into the operation of all literature" (18). Edwards is correct to point out that game theory emphasizes rules, choices, unambiguous consequences, and so on, and that literature (as one form of aesthetic activity) is playful. It is equally true that the particular disciplines that Edwards singles out for attention—"economics, political science, and resource and military planning"—are not notable for their playfulness. That those disciplines use formal mathematical game theory—really the theory of choices, the science of decision-making—does not make them into games. This is why the disciplines listed by Edwards are not usually considered games, though they use game theory: they tend not to play, preferring unplayful and exact precision instead; as much as possible, they try to eliminate play, in fact. To say, on the other hand, that literature is *fundamentally* playful does not imply that literature is *therefore not* a game: games do not stop being games merely because they emphasize what Caillois calls "paidia." While play is certainly "the more expansive term" of play and game, that does not mean that game is "subsumed" by play; on the contrary, if Bernard Suits's definition of game (of which more below) has any validity, true game *assumes* play. It could be said, in fact, that game subsumes play in at least this respect: game can be defined as limited, or as structured, play.

In his *In Palamedes' Shadow*, Robert Wilson writes:

> For the purposes of the immediate argument, some of the easier senses in which a literary text might be a game have been excluded:

> that writers play games with themselves when they adopt certain conventions, that they play games with themselves when they set goals or impose limitations upon themselves, that they play games with readers who must discover the conventions that writers have chosen and avoid the snares they frequently set, that readers play games with texts by imposing interpretative rules on them, that readers play games with writers by outsmarting them and by making of their texts what they had never imagined, are all notions, neither indefensible nor unreasonable, that have had many theoretical defenders in recent years. (94–95)

I do not wish to make use here of any of "the easier senses in which a literary text might be a game." One can play games within or around virtually any discipline, without thereby making that discipline itself into a game: writers can make a game of writing novels if they so wish, without thereby making novels into games (what writers do alone in front of their word processors is entirely up to writers); readers can make games of reading novels, also without thereby making the novels that they are reading into games (what readers do when they are alone belongs in the same category as what writers do in the same state). In their turn, novels can make use of scientific facts or of history without making those things any more, or less, "factual" or "historical." One can imagine playing a game based on the principles of geometry or on the laws of algebra, without concluding that therefore geometry is a game, or that algebra is anything but the elegant mathematical system that it is: neither becomes a game merely because its premises are used in a game situation.

Bernard Suits has written extensively about the investigation of games in literary analysis. He suggests several steps to use in conducting such analyses:

> (1) It is necessary at the outset to have a reasonably clear idea of what a game is.... (2) It is then necessary to specify which particular literary process one is talking about in applying a game-analysis to it. Is it (a) the process of literary creation, (b) the literary work itself (or some part or aspect of it), or (c) the response of readers to the work? (3) It is further necessary to specify, in applying a game-analysis to one or all of these processes, whether a game is being constructed, played, or viewed.... (4) Finally, it is necessary to examine literary forms (and possibly individual works) *one at a time*, for the strategy seeks to

> discover what in literature is genuinely game-like and what is not. ("The Detective Story" 215)

Several of Wilson's "easier senses" correspond essentially to Suits's points (2a) and (2c); as I have mentioned, I do not wish to discuss them here. I do, however, intend to follow Suits's suggested techniques for detecting games in literature. First, then, I shall ask this question: "What is a Game?"

A. *Negative Definitions*

As a first step towards defining what it is I mean by "game" as that term applies to postmodernist literature, I shall enumerate several of the ways in which the literature-game analogy will *not* be used in this book.

I shall not maintain that novels are puzzles, for instance. This is the strategy, in essence, that Suits uses in his discussion of detective novels.[10] John Fowles discounts that theory as it pertains to non-detective fiction rather eloquently in his foreword to *The Magus*: "Novels, even much more lucidly conceived and controlled ones than this, are not like crossword puzzles, with one unique set of correct answers behind the clues.... Its meaning is whatever reaction it provokes in the reader, and so far as I am concerned there is no 'right' reaction" (9–10).[11]

It is clear, too, I think, that just because the subject matter of a given novel is a game, that does not make the novel itself into that, or any other, game. Thus I can discount as unhelpful in establishing the gamefulness of fiction such novels as William Kennedy's *Billy Phelan's Greatest Game*, which is, in part, based on a game of bowling; or Robert Coover's *Whatever Happened to Gloomy Gus of the Chicago Bears?*, a novel about football, or *The Universal Baseball Association*, which is a novel about a table-top dice and board game of baseball, but is not that board game itself, much less a game of baseball itself.[12] Equally, Italo Calvino's *The Castle of Crossed Destinies* cannot be confused with an actual deck of tarot cards. I have already mentioned Lewis Carroll's *Alice* books and their connections to chess games and chess problems; Nabokov is also very fond of chess, and uses the game in numerous novels and collections, but none of those novels or collections *is* a chess game. Many novels of manners (Jane Austen's *Pride and Prejudice*, for instance) are full of characters who play games, from various card games to backgammon, yet *Pride and Prejudice* is neither card game nor backgammon match.

There are also games in literature of a slightly different order from any of these. Walter Abish's *Alphabetical Africa*, for instance, is a curious technical exercise. All of the words in the first chapter of that novel begin with the letter A; the first word in the second chapter begins with B, and all the words in that chapter begin with either A or B; the first word in the third chapter begins with C; and so on through to Z; subsequent chapters work back to A again. Abish may well have played a game with himself in setting up the very formal constraints of the novel before he actually wrote it. In fact, it is obvious that those constraints must have been formed before the novel was begun. *Alphabetical Africa* appears, in short, to be a game in sense (2a) of Suits's schema in "The Detective Story" (215), but that is a sense which does not help my argument in this book.

Alternatively, B.S. Johnson's *The Unfortunates* is composed of twenty-seven separate sections (essentially they are unbound chapters) and comes in a box. The inside front cover carries this note:

> Apart from the first and last sections (which are marked as such) the other twenty-five sections are intended to be read in random order. If readers prefer not to accept the random order in which they receive the novel, then they may re-arrange the sections into any other random order before reading.

The Unfortunates is also an account of a game (a soccer match). The novel-in-a-box, or the "shuffle novel," can be seen as a prototype of hyperfiction (of which more later), but in this case, the narrative resolves itself fairly readily into a first-person psychological novel: obviously, the order of exposition changes somewhat when the sections are shuffled, but the novel's story-line stays substantially the same. If there were an exact preferred order in which *The Unfortunates* needed to be read, and that order were to be kept a secret by the author and publisher (as master game-wrights), and that order needed to be discovered for the novel to be read or to be readable, a case could possibly be made for its being a game of sorts; but the novel is, it says, to be read in a random order, and although the box that contains *The Unfortunates* could be said to contain several novels in one, it is difficult to see how it is a game.

Of possibly more promise is the novel that was "edited" by Margaret Drabble and B.S. Johnson, *London Consequences*, whose last two pages read this way:

> Each chapter of this novel was written by a different author; the first and last are the joint work of the editors, Margaret Drabble and B.S. Johnson, but the remainder are anonymous. The Greater London Arts Association is offering a prize of £100 for the correct identification of the authors of these sections....
>
> Who Wrote What? Competition Rules
>
> 1. The closing date for receipt of entries is 11 August 1972.
> 2. All entries must be on the form provided, and no others will be accepted.
> 3. Competitors may enter any number of times, providing that each entry is on the correct entry form.
> 4. Winners will be notified by post and the results will be announced in the October issue of *Greater London Arts*, the Association's monthly newsletter.
> 5. Voluntary officers and employees of the Greater London Arts Association will not be eligible to compete.
> 6. In the event of a tie, the prize will be divided amongst the winners.
> 7. In the event of there being no correct entries, the judges may award the prize at their discretion.
> 8. The judges' decision shall be final.

This is clearly a game. It is a puzzle with a single correct answer; it is a contest that offers a substantial reward for its solution. However, although the game is contained in a novel, and although it concerns a novel, it is not a demonstration, either that this novel, *London Consequences*, is itself a game, though it contains a game; or that novels more generally are games, although they can, as I have already noted, contain games, or be the subject matter of games.[13]

Julio Cortázar's *Hopscotch* is sometimes cited as a novel that requires one to play a game in order to read it. It begins:

> In its own way, this book contains many books, but two books above all. The first can be read in a normal fashion and it ends with Chapter 56, at the close of which there are three garish little stars which stand for the words The End. Consequently, the reader may ignore what follows with a clean conscience. The second should be read by beginning with Chapter 73 and then following the sequence indicated at the end of each chapter. In case of confusion or forgetfulness, one need only consult the following list.... (5)

There are several things that need to be noted about *Hopscotch*: first, chapter 55 has no "following" number; it is only read once, in the first ("normal") version, directly between chapters 54 and 56. It does not appear in the second version at all. Chapter 131, on the other hand, is read twice in the second version. Aside from those small differences, however, the novel does not change very much from one version to another. Chapters 1–56 are chronological, or at least arranged conventionally, according to the narrative time line of the novel, and whether one reads the novel in the first way or in the second, the chapters are read in that order; the second set of chapters, although they are numbered in an apparently random order, are interpolated between chapters from the first "normal" reading, yet leaving the order of those first chapters intact. The chapters which make up the second set, moreover, are *expendable* chapters, considered from a narrative perspective: they range from quotations from Claude Lévi-Strauss, for example, to bits of letters. They are, in short, additions to a stable narrative; while they may appear to make the novel into a rather haphazard construction of readings deviating wildly from the first, "normal," reading, they do nothing of the sort. The "second book" is really only an amplified version of the "first book."[14] Thus if *Hopscotch* is a game, it is merely a game of misdirection, wherein readers are led to expect something which the novel does not deliver (namely, the promise of "many books, but two books above all" in one).

Finally, novels which address the reader directly are sometimes considered to be game-like, although I am rather more inclined to class this strategy as a blurring of boundaries and roles—as play, in short, rather than game. The famous opening lines of Italo Calvino's *If on a Winter's Night a Traveller* ("You are about to begin reading Italo Calvino's new novel, *If on a Winter's Night a Traveller*...") comprise an example of this sort of direct reference,[15] as does the quiz which appears at the end of Part I of Donald Barthelme's *Snow White*:

> 1. Do you like the story so far? Yes ( ) No ( ) ....
>
> 15. In your opinion, should human beings have more shoulders? ( ) Two sets of shoulders? ( ) Three? ( ) (82–83)

B. Positive Definitions

Newton Garver writes of game:

> What is important is this: if an activity is identified as a game, it must be so identified upon the basis of a structure in the activity such as might be described by rule formations, that is, by explicitly stating what the players may or must or cannot do at certain times or in certain circumstances. If an activity is not implicitly or explicitly governed by rules—that is, if there are not regularities discernible in it, or if the regularities in it are accidental rather than prescribed—it cannot be a game. Games, therefore, are constituted or defined by the rules according to which they are played (although the identity conditions are not so strict that every rule change need result in a new game). (232)

In short, according to Garver, whereas play is defined by its freedom, its anarchy, its openness, its resistance to closure, game is defined by its rules, constraints, limits, and borders.

The primary definition of game that I wish to use in this study fundamentally agrees with Garver's definition, but adds some refinements. It is taken from Bernard Suits's *The Grasshopper: Games, Life and Utopia*. In that useful work, Suits defines games in this way:

> To play a game is to attempt to achieve a specific state of affairs [prelusory goal], using only means permitted by rules [lusory means], where the rules prohibit use of more efficient in favour of less efficient means [constitutive rules], and where the rules are accepted just because they make possible such activity [lusory attitude]. (41)

This is also the definition of game that Robert Wilson uses in his *In Palamedes' Shadow*. Noting, with Suits, that "there is a good deal of loose talk about games these days,"[16] Wilson goes on to discuss the applicability and usefulness of game, rigorously defined, as a model for literature and literary study. He concludes that the concept of game, as defined by Suits, is not usually useful in discussions of literature: most scholars who discuss game and literature do so by conflating terms and blurring distinctions.[17] Wilson's conclusion is based on a crucial "distinction between convention and rule, between a set of flexible assumptions and a dissimilar set of inflexible assumptions" (103).

The more precisely a game can be defined by its rules, the more precise an example of "Game" it becomes. In chess, for example, little beyond a working knowledge of the rules is necessary to be able to play the game (although there is considerably more than this involved in playing the game with distinction). This elegant clarity is one reason that chess is used so widely in literary studies as an exemplary game. Chess adheres to its rules so exactly, and there is so little involved in the game outside of the making of moves that adhere to the rules, that it is a simple matter to record complete chess games for future reference, simply by recording moves. Of course, not all games are as closely circumscribed by their rules as chess is. For example, as a game becomes more physical than chess, it becomes increasingly difficult to record that game in the ways that chess can be recorded. How does one record the "moves" of a major league baseball game? Even if one had all the statistics of a particular game at hand, how would one go about recording the grace, skill, and coordination of, say, Roberto Alomar ranging far to his left for a sharply hit ground ball and flipping a throw to get the runner by half a step? "Ground out, 4–3," just doesn't do justice to what happened, even though it records precisely what did happen. The same consideration obtains in other sports. In soccer, for instance, the statistics do not capture the skill of a goal by Pele on one of his patented "bicycle" shots. No numbers or statistics express the matchless grace of Michael Jordan with a basketball or of Wayne Gretzky with a hockey puck. On the other hand, Q X B ch tells the chessplayer everything that can be known about that move in that game. The chess move has its own elegance about it, contains its own elegance in itself; however, it is the *execution* of the move that lends elegance to the move in a professional sports game.

There is no rule demanding grace in ice hockey, yet the grace of a highly skilled centre is part of that game. The skill of the chessplayer is demonstrated by her masterful use of the moves required or allowed by the rules and the constraints of the board. Her skill is demonstrated by what she does. The skill of a basketball player like Jordan, on the other hand, is demonstrated not only by what he does, but by how he does it, as well. Thus, although it is true that rules define a game, that does not mean that a recitation of the rules and of the way the rules have been followed in a game can capture the whole of a game. If postmodernist literature can be seen as game-like, it will probably be a quite complex game, involving not only rules but many conventions, and many matters of interpretation.

Rules in the context of a game define the game. They limit it, spatially, temporally, logistically. Runs do not count in baseball, for instance, after three outs have been recorded in an inning; goals do not count in hockey after time has expired; or in soccer if there is a player who has gone offside. Rules not only limit the game, however: they also make it possible. Trotting around a ninety-foot square only counts as a home-run in the context of a baseball game; otherwise, it is just trotting. The rules make basketball possible, even while they limit and constrain the play: a ball going through a hoop is only a ball through a hoop, until the context of a basketball game makes it a turnaround jumper from three-point range. Alternatively, imagine a labyrinth without walls or limits of any kind: even the most complicated labyrinth becomes a field, or a parking lot, or a large empty building. The labyrinth requires limits—walls, hedges—in order to be a labyrinth. In negotiating a labyrinth, one is usually allowed to range back and forth and play in the corridors interminably; but one becomes a labyrinth spoilsport if one gets to the end or the heart or the other side of the labyrinth by stepping through the hedgerows or by climbing over walls or by going around the outside of the structure. In the same way, a game without rules not only ceases to limit actions, but at the same time ceases to be a game.

C. Rules

If rules are really at the heart of what makes a game a game, then rules also need to be defined.

In an often-cited paper entitled "Two Concepts of Rules," John Rawls discusses the nature of rules. Rawls claims that there are two main kinds of rules. The first is what he calls "summary" rules. Under this "conception,"

> ... it will happen that in cases of certain kinds the same decision will be made either by the same person at different times or by different persons at the same time. If a case occurs frequently enough one supposes that a rule is formulated to cover that sort of case. I have called this the summary view because rules are pictured as summaries of past decisions arrived at by the *direct* application of the utilitarian principle to particular cases. Rules are regarded as reports that cases of a certain sort have been found on *other* grounds to be properly decided in a certain way (although, of course, they do not *say* this). (19)

These are rules of conduct, rules of decision-making. In courts of law, for example, they are called precedents; in other, less formal situations, they are codes of behaviour, ways of acting. "The point of having rules derives from the fact that similar cases tend to recur and that one can decide cases more quickly if one records past decisions in the form of rules" (22). These rules are summaries, guides, maps to behaviour. It is possible to see this first kind of "rule" coming into effect in game situations. Consider the baseball player who notices a certain movement on the part of the pitcher, and concludes that that movement often signifies that it is safe to steal. Alternately, consider the poker player who notices that her opponent has a noticeable tic when she is holding a good hand. The player's "rule" for base stealing or for laying bets is of a different order from the rule that dictates a limit of three strikes on a batter, or of five cards in a hand at a time. The rules allowing only three strikes or five cards are examples of Rawls's second kind of rule.

Under Rawls's second conception, the direction of authority, or of the rule-making, changes. Rather than the rule following and describing or summarizing a practice, under the second conception, the rule comes first, and defines the practice:

> In contrast with the summary view, the rules of practices are logically prior to particular cases. This is so because there cannot be a particular case of an action falling under a rule of a practice unless there is the practice.... One may illustrate the point from the game of baseball. Many of the actions one performs in a game of baseball one can do by oneself or with others whether there is the game or not. For example, one can throw a ball, run, or swing a peculiarly shaped piece of wood. But one cannot steal a base, or strike out, or draw a walk, or make an error, or balk.... (25)

As Rawls points out, under the first conception of rules, one can, and indeed one often does, cite the rule while petitioning for an exception. However, Rawls writes that in the context of baseball's rule of three strikes on a batter (which falls under the second conception), to claim that "it would be best on the whole" that the rule not be followed (26), and that a given batter should in this instance be given a fourth or fifth strike, is absurd.

In games, the rule defines possible or permitted or desirable actions: "If one wants to play a game, one doesn't treat the rules of the game as guides to what is best in particular cases" (26). In short,

> The practice view leads to an entirely different conception of the authority which each person has to decide on the propriety of following a rule in particular cases. To engage in a practice, to perform those actions specified by a practice, means to follow the appropriate rules. If one wants to do an action which a certain practice specifies then there is no way to do it except to follow the rules which define it. (26)

Under Rawls's system, it is this second kind of rules—rules of practice—that govern games, that comprise what Suits calls the "constitutive rules" of games.

Max Black has also written about rules; in particular, he addresses rules in a chapter in *Models and Metaphors* devoted to "The Analysis of Rules." Black distinguishes four kinds of rules, as follows: (1) rules of regulation; (2) rules of instruction, or directions; (3) rules of precept, or maxims; (4) rules of principle, or general truths (109–13). Many of these kinds of rules are self-evident; most of them are not pertinent to the discussion at hand. What is of interest, however, is Black's first kind of rule, the rule of regulation:

> It is sometimes said that certain games (or other systematic activities) are "constituted" by a set of rules.... To say that chess is constituted by the rules of chess is to say that a man who failed to "heed" the rules *would not count* as playing the game. Or to put the matter in a slightly different way, our notion of the game of chess is the notion of something played *according to* certain rules. (123)

In essence, Black agrees with Rawls; his "rules of regulation" are essentially synonymous to Rawls's "rules of practice." In both cases, rules are constructed to dictate and structure what is required, permitted, or forbidden in a certain practice. "That is to say," writes Black, "that when we think of this kind of rule, we are, more or less clearly, thinking of cases where somebody has authority to tell us what to do" (119).

Finally, Black distinguishes between cases in which regulations apply to, but do not "constitute," the activity in question:

> Games of skill are played for the sake of winning *according to the rules* (if all you wanted was a touchdown, why not shoot the opposing team?); whereas the aim of parking is to get one's car off the street in a safe place, the aim of respecting the relevant laws in so doing being subordinate (and, as it were, "external"). Think of the absurdity of arriving in a remote village where there are no parking regulations at

> all and saying "Too bad—it's logically impossible for me to park here." (124)

Parking regulations do not constitute parking, though they govern parking in most circumstances. Basketball's basic constitutive rules, however, define, limit, and thereby constitute the game of basketball, even though there are other rules, practices, conventions, and skills involved in basketball besides the following of those rules.

D. Conventions: A Parable

The first important move that Robert Wilson makes in his discussion of the "Game/Text Analogy" in *In Palamedes' Shadow* is to relay a parable about two people canoeing on Lake Louise in the Canadian Rocky Mountains. The story has it that "a young couple in a canoe about one hundred yards out on the lake" "abruptly" become the object of derision of a laughing, gesticulating crowd because

> They were both energetically occupied in paddling ... [but] the two were facing each other in the canoe and thus all their vigorous effort to move forward resulted only in a circular movement. They were spinning in tight circles and spinning more quickly the more determinedly they attempted to paddle forward. One has only to imagine the scene to perceive that the young man and the young woman have done something wrong. Clearly, two people do not face each other in a canoe if they wish to paddle together, cooperatively, in a forward direction. Of course, they have acted correctly if they *want* to move in circles, but since one assumes that circular motion is not the normal objective of two-handed canoe paddling, one is justified in inferring that they have done something wrong. (83)

While Wilson concedes that it is possible that the two paddlers may have wanted to move in circles, he dismisses this possibility as unlikely because "circular motion is not the normal objective of two-handed canoe paddling." Wilson uses this parable to begin to illustrate the distinctions between conventions and rules: although it is the custom—the convention—that the two paddlers face the same way while they paddle, it is not a constitutive rule of sitting in a canoe or of canoe paddling that they do so. As Wilson notes, this customary seating arrangement is not a rule, but a convention, and a sensible one to use if the canoeists wish to make straight-ahead progress. In facing each other, of course, the canoeists have not broken a rule: no "rule" exists that proscribes face-to-faceness in a canoe. "But a simpler, less exacting stipulation than a rule" has been breached: a convention (83).

There are, however, two possible interpretations of the case of the two canoeists. The first is this: the two are incompetent canoeists (converted Humean row-boatists, perhaps), unaware of both canoeing conventions and the laws of physics. They are trying to move unobtrusively in a forward direction across the lake, and they are becoming increasingly frustrated as, despite their best efforts, they succeed only in spinning in ever-tightening circles. This is the approximate interpretation that Wilson uses to good effect to demonstrate the difference between rule and convention.

The second interpretation is that they are in fact competent canoeists; that they are deliberately defying canoeing conventions to sit facing each other; that they are canoeing "abnormally" for some reason of their own; that they are, in short, playing a game, and are doing so despite the "normal objective[s] of two-handed canoe-paddling." This interpretation in no way questions the distinction between rule and convention (a distinction which was, after all, the point of the parable in the first place); it is designed, rather, to demonstrate that the boundary between rules and conventions may sometimes waver. Rules may, on occasion, look like broken conventions.

Three things need to be noted about the canoeing scenario as presented. First, the parable specifies that the laughter of the spectators erupted "abruptly." Second, the canoeists were, when the laughter erupted, "about one hundred yards out on the lake." Third, the "vigorous effort" of the canoeists "resulted only in a circular movement" that increased as the efforts of the canoeists intensified. That the laughter erupted abruptly after the canoeists were already one hundred yards out on the lake suggests that they arrived there unobtrusively. In fact, if the canoeists were truly ignorant of "normal" canoeing procedures, it is likely that they would be spinning in frustrated, rocking circles next to the pier or the dock and receiving much-needed instructions in canoeing technique and seating conventions from the spectators on shore. Instead, the canoeists have arrived at a spot well away from shore, where they can do neither themselves nor anyone else harm or inconvenience; once there, they (abruptly?) begin to spin. Their location itself suggests that they have canoeing competence, and that they have used their competence to get to where they were when the laughter started. Finally, they must be very dull folks indeed if they do not discern that the ever-increasing vigour of their efforts results not in progress across the lake, but the ever-more-profound thwarting of their progress.

Wilson's parable, while fulfilling well its original purpose—to illustrate the difference between rule and convention—also potentially illustrates a kind of simple game, one that conforms to all of Suits's constraints, including having at least one constitutive rule. The game has a conceivable "prelusory goal": "to spin until we are dizzy," perhaps, or "to paddle until we are tired." The "lusory means" of the game could be "to spin our canoe using only these paddles" (a conceivable summer-camp variation could be "to paddle the canoe using only our hands"; this option is not taken up by the Lake Louise paddlers). A "constitutive rule" could be something as simple as "we must paddle in circles like this until one of us must stop due to exhaustion," or "we must paddle in circles like this as long as possible, and nothing but the paddles may touch the water." Indeed, a perfectly plausible constitutive rule could be this: "we must sit facing each other in the canoe and paddle as hard as we can for as long as we can." The "lusory attitude" by now should be clear: "we are doing this purely because we want to; we do not care about crossing the lake; we do not care about canoeing conventions; we are out to have fun—we are playing a game." Indeed, it is worth noting that several of the possibilities of gamefulness in this example evaporate if the two canoeists sit facing the same way, as canoeists engaged in normal, straight-ahead canoeing would and should sit. The game-playing canoeists, by contrast, are not engaged in transportation, but something quite different, even though the equipment they are using is conventionally used by canoeists as a pleasurable mode of transportation on beautiful lakes in the Canadian Rockies.

The paddling game as I have sketched it displays many of the characteristics of "normal" canoe paddling: it involves two persons, a canoe, a lake, two paddles. Canoe paddling, however, as Wilson has demonstrated, is not in itself a game, but a convention-laden activity: it is potentially a mode of transportation; a form of exercise; perhaps even (in the case of the young couple on Lake Louise, for instance) an elaborate form of courting ritual. Game playing and canoe paddling intersect in that space where rules overlie and govern—even dictate—conventions (as in the stipulation that "we must sit facing each other"). Even if the constitutive rule of the canoe-paddling game is to "sit facing each other in the canoe and paddle as hard as we can for as long as we can," certain conventions—many of the ones applicable to straight-ahead canoe paddling—are still imperative for this game to work: the paddles are obviously held by the handles, with the blades touching the water, for instance; and the paddles are dipped into and stroked through the water, and not waved about one's companion's head. Furthermore, whether the scene is read as an example of transportation frustrated or of a game played has less to do with the evidence available to the reader (which remains constant) than it has to do with the interpretation of that evidence. More contextual evidence might help us to decide which interpretation is more complete: knowing for sure how the two canoeists arrived one hundred yards out in the lake could help us to infer their degree of competence, for instance, as might the ability to eavesdrop on their conversation or to see the expressions on their faces as they strain at their paddles. Whether the two canoeists are frustrated travellers or ludic paddlers is, no doubt, clear to them; whether they are actually playing a game is not dependent on a watcher's interpretation. However, for us watchers, lacking complete evidence, and being unable to interview them, their actions are evidence only, the stuff of which interpretation is made.

Wilson continues: "The fact that literary texts are self-enclosed, separable from other activities, possess their identifiable text-specific assumptions (which is their enabling legislation), and are, or seem to be, autotelic has promoted the analogy with games" (85). However, he notes, the "enabling legislation" leading to game and that leading to literature leaves a gap so wide that it "suggests no fruitful analogies" (85). Wilson wishes to demarcate the lines between game and literature, to point out that game and literature are not equivalent, that they operate on different principles. I have already mentioned that his argument hinges on a distinction between rule and convention, a distinction that needs to be clear:

> If rule and convention are synonymous, if they do point to the same cultural phenomena, it will be necessary to grant that *axiom* is equivalent to *rule* and thus to *convention*. Literary criticism could then inherit, joyfully or tearfully, an entire series of interchangeable terms, *axiom*, *law*, *rule*, *convention*, *direction*, and *supposition*, each indicating the range of diverse assumptions that constitute semiotic and cultural acts. Doing geometry, playing chess, writing a narrative, baking a casserole, paddling a canoe, or merely supposing, as in a daydream, a possible state of affairs would all appear to be equivalent because they would all imply similar constitutive demands. (87)

While I acknowledge the importance of all of those distinctions, I would suggest nonetheless that a game could be constructed within virtually any one of those activities, even though they are not games to begin with, and are not themselves only, or even primarily, games. This can be done by adding rules that shape and delimit the conventions that are already there.[18] Just as the most obvious and important aspect of the behaviour of the canoeists was that they were sitting facing each other, and so either ignoring or transforming a convention of canoe paddling (perhaps for the sake of a rule), so can the activities of "writing," or "baking," or even "supposing" be transformed into games by the addition of one or more appropriate rules and attitudes.

Indeed, working as a tutor one summer I made up a game for my young charges out of the routine of learning simple equations: doing simple mathematics in itself is certainly not a game (though it may be a job, conceivably a hobby, most likely a child's onerous task), but with the addition of simple rules, a lusory attitude, and perhaps a brother or sister with whom to compete, a game can be played inside mathematics, a game that shares many conventions with the straightforward doing of mathematical equations. In this game, 2+2 does not magically become anything other than 4 (indeed, the point is that the link between 2+2 and 4 be formed firmly): the essential activity of mathematics does not change when a task involving mathematics becomes a game. A popular children's game is "Button Button So High," generally played in my household with no "real" button, but based entirely on "supposing ... a possible state of affairs" that is gradually revealed through guessing to the other game-player. That idle supposing alone is not a game is, or should be, evident; however, the addition of rules and a gaming attitude can transform supposing into game.

Wilson then moves to a more concrete example, drawing distinctions between "chess" (whether Indian or European), with its well-defined and very formalized rules, and "pastoral," with its loosely defined, floating conventions. This example is carefully chosen: it makes the point regarding rule and convention (as it is specifically chosen to do) well, although it is (deliberately, from the first) a false comparison. Chess is a game that is almost entirely stripped of convention. Indeed, almost the only convention left in chess has to do with the board and the chessmen.[19] It is commonplace for critics to point out that chess remains chess whether played with common kitchen vegetables on a checked table cloth, or with an elaborately carved antique ivory and ebony set of chess pieces on an equally elaborate board, or with no physical pieces or board at all. The conventional parts of the chess game—the pieces, the board—do not make possible the game of chess: the rules of chess do. Pastoral, on the other hand, is (only, ever) a set of conventions that crosses genres, types, modes. Thus there are pastoral romances, pastoral elegies, pastoral odes, pastoral comedies, and so on. Where chess is defined by rules, which give it its form, and which (generally speaking) do not change, pastoral is defined by conventions: that is, its content, which can change with context, mode, mood, etc.

A comparison from literature that might have a greater resemblance to chess than "pastoral" is the sonnet.[20] Sonnets are identified as sonnets not by their content (which is strictly conventional, and, conventionally, secular love), but by their form. One learns the rules of sonnets in very much the way one learns the rules of a game: the rules that govern both activities are "abstract, easy to formulate, descriptive as well as prescriptive, predictive, and inflexible" (94).

A sonnet by Donne, then, might be about religious devotion; one by Wordsworth might be about social unrest in early nineteenth-century England; one by Yeats might be about the occupation of Ireland by the English. One can imagine a postmodernist sonnet (by John Ashbery, say) about the sonnet form itself. In all cases, however, the sonnet is not defined in terms of its content (which is liable to be traceable to part of one conventional mode or another: romantic love, social protest, pastoral), but according to its form.

In addition, the rules of sonnets are stylized enough to permit distinctions to be made (along the lines of the distinctions between European and Indian chess) among kinds and variations of the sonnet form: Petrarchan, Shakespearean, and Spenserian sonnets, for instance. Just as a chess game without the knight's move becomes a very different game, so a poem with only thirteen or with fifteen lines ceases to be a sonnet (even if that poem is written in iambic pentameter and is about secular love).[21] Just as the rule governing a pawn's first move delineates the boundary between Indian and European chess, while the game remains recognizably chess, so variations in rhyme scheme demarcate whether a particular sonnet is a Petrarchan, Shakespearean, or Spenserian one. In short, rules are likely to have more to do with form than with content; conventions, more with content than form.

An objection could be made here that iambic pentameter is important to the sonnet, is part of its definition,[22] yet there are poems that are irrefutably sonnets that do not conform rigidly to the strictures of iambic pentameter. The use of iambic pentameter seems more central to sonnet-hood than whatever is described by mere convention,[23] runs the argument, yet if it were a constitutive rule that is in fact as unforgiving as constitutive rules have behaved so far, likely every sonnet ever written, and certainly all of Shakespeare's, would be disqualified from sonnet-hood for breaking the rules.[24] If we were rigid about the formal shape of the sonnet, we would, perhaps, need a category of poetry that included the "near-sonnet." Further, runs this argument, even though the sonnet is usually discussed as an iambic pentameter form, talk about iambic pentameter in sonnets often insists upon the "pentameter" much more than the "iamb"; much discussion of iambs, in fact, centres on variations away from that norm. It is a part of literary study to note differences and distinctions: countless classrooms full of first-year students have learned that poets sometimes use variations on the iamb for deliberate effect.

Rules that are not essential to the game, Suits calls "rules of skill." Suits gives as "examples ... the familiar injunctions to keep your eye on the ball, to refrain from trumping your partner's ace, and the like" (*The Grasshopper* 37–38). Two things seem important to note about "rules of skill." First, they are similar to Wilson's "conventions," except they appear to have even less determining force than conventions. They appear, in fact, to be entirely concerned with whether a player is playing well or badly, and not at all concerned with delimiting the game. For the purpose of this discussion, it may be more useful to call them neither rule nor convention, but merely advice. Second, the use of the iamb seems not so much a rule of skill of the sonnet as a constitutive rule of the sonnet that is bent or broken from time to time for effect. The rule of skill governing the iamb in the sonnet, in fact, may be an injunction to "vary the iambic pattern from time to time, or your poem may become monotonous." That sonnets are fourteen-line single-stanza lyric poems with certain pre-set rhyme schemes written generally in iambic pentameter seems true: this statement seems to summarize the constitutive rules of the sonnet. However, that some of those constitutive rules are broken from time to time without the sonnet becoming something other than a sonnet seems also to be true: much criticism and teaching have been devoted to the effect that a trochee can have at the end of an iambic pentameter line.

Rules governing iambic pentameter; rules governing sonnet rhyme schemes (Petrarchan-Shakespearean-Spenserian); rules governing the pawn's moves (Indian-European chess); professional sports rules, such as the one that governs whether a pitcher hits for himself or a "designated hitter" hits for him (National vs. American League baseball): all these seem to be constitutive rules that, for one reason or another, are of a different order of magnitude from other constitutive rules (say, "sonnets" with thirteen lines, "chess" without a knight, "baseball" without a pitcher). Baseball, and sports generally, seem to have many rules that operate in this manner. Some fans would not notice the difference in a baseball game wherein the infield fly rule was not enforced, although they would certainly notice the difference if only eight players appeared on the diamond, and no one took up a position on the pitcher's mound, or if the game were suddenly to be played according to a time clock rather than according to innings as it is currently played. There seem to be differences in the relative importance attached to particular constitutive rules. Those differences, further, seem to be a matter of situation and interpretation. It may be relatively unimportant, for instance, to point out the variations in the first move of a pawn in chess, unless one wishes specifically to discuss the variations between Indian and European chess, though that distinction marks a (slightly) different game. Of course, both players involved in a particular chess game need to know what set of rules, exactly, is being used in that game. Equally, the designated hitter may not be a factor in any discussion of baseball that does not mention the differences between the two professional major leagues of baseball, yet the DH is probably the single major difference in the games the two leagues play. And neither of those rule changes changes its respective game from being recognizably chess, or baseball. In short, as Allen Thiher notes, "There is ... more to a game than merely following rules" (21); conventions are also important.

E. Gamefulness and Literary Postmodernism

To close this chapter, let me summarize my position with regard to Suits's suggestions in "The Detective Story" for detecting games in literature.

First, my definition of game remains the one from *The Grasshopper*:

> To play a game is to attempt to achieve a specific state of affairs [prelusory goal], using only means permitted by rules [lusory means], where the rules prohibit use of more efficient in favour of less efficient means [constitutive rules], and where the rules are accepted just because they make possible such activity [lusory attitude]. (41)

The prelusory goal of reading a novel is connected, of course, to various motivations for reading generally: escape; entertainment; relaxation; aesthetic bliss; the gathering of information; and so on. In short, the reading of a novel is usually done fundamentally for its own sake. One reads novels just to do so; this reasoning is similar to that which takes place when the reasons for playing games are listed. Similar claims can also be proffered for Suits's criteria regarding "lusory means" (that is, reading itself, presumably from the beginning to the end of a novel) and "lusory attitude" (fun; playfulness). The crux of whether the game-literature analogy holds, for Wilson, and for other theorists, has to do not with those three elements, however, but with whether there is at least one constitutive rule that governs novels and novel reading. Without at least one rule, there cannot be a game; without a game, it is difficult to imagine "gamefulness" as a useful concept.

The rule that I wish to suggest regarding the postmodern novel is simple, abstract, easy to articulate, and easy to understand. It is a rule which defines postmodernism contextually by setting the limits of a given novel. The consequences of ignoring the rule or of getting it wrong is that the novel in question will change from being postmodernism to being something else: realism, perhaps, or a romance, and very often a failed example of one of those modes.[25] The specific limits of a particular novel may shift, depending on the novel in question, although the principle that there are limits that are constructed in and by the postmodern novel as part of a self-reflexive self-definition does not change. Thus the limit of postmodernism, the edge that marks John Barth's *The Last Voyage of Somebody the Sailor*, for example, as postmodernism, that defines it for what it is, is inscribed primarily by its intertextuality, and by other literature and story more generally; in Louise Erdrich's *Tracks*, the limits are drawn by storytelling and historical land claims and racial and cultural barriers.

The rule is this: when one is reading a postmodernist novel, one must know that one is doing so, and read "postmodernly." That is, on one hand, one must read playfully, not expecting closure, not expecting traditional novelistic elements to be the primary mode of exposition; and, on the other, one must expect and look for the limits, the borders, the signs of the parameters of the discourse, and the limits of the fiction. The postmodernist novel is likely to be bounded by one or more elements that in another context would be considered conventions, but in literary postmodernism, will harden into a rule, or at least into a limit, as a labyrinth has walls. The novel in question will play; it will resist closure; its meanings will be multiple, undecided, and undecidable; it will be charged with a restless, dancing, playful energy. It will acknowledge that it does and will always play, and so partake of postmodernism. And it will have limits; it will have structure; it will construct itself as a novelistic artefact, and so partake of the tradition of the novel as well. It is liable to be inventively playful with language, conventions, and the lines between history and literature or between fact and fiction.

The limits of given pieces of postmodernist fiction are not pre-given; they must be constructed as they are read. As Lyotard notes, because the "master narratives" that used to supply the limits to novelistic (and other) discourse(s) have broken down, because the many conventions that used to obtain have been destroyed, postmodernism is charged with the task of setting its own limits, making its own rules (*The Postmodern Condition* 83); the postmodern novel is always therefore a rule maker and a playful convention trespasser, at one and the same time.

As far as Suits's other suggestions are concerned: first, I expect the rules of gamefulness in postmodernism are to be found in the literary work itself, and are to be found there with the cooperation of readers (meanings arise and are multiplied, after all, in the process of reading), and with the gleeful participation of authors (though they are dead, according to Barthes, or are corrupt "plagiarists and paraphrasers and brain pickers and mocking birds," according to William Gass ["The Death of the Author" 24]). Second, I think that the game in question in literary postmodernism is very often constructed, played, *and* viewed; I am not sure that in the context of postmodernism the three can be entirely separated. And third, I agree with Suits that it is necessary to examine literary forms one at a time; my discussion, in fact, specifically entails examining individual novels one at a time. It is to that examination that I turn next.

3. Conclusion

In this chapter, I hope I have shown at least these things: first, that the boundaries between conventions and rules, though important, are not as precise in all situations as they are, for example, in chess. In fact, I hope that I have established that the line between rule and convention can been seen as a continuum, not as a knife-edged border. Second, that the boundaries between game and non-game can waver (and the same evidence can be mined for each), and that sometimes games are only discernable when one has access to such murky subjects as the attitudes of the players. Third, that there is a sense in which gamefulness and the detection of gamefulness are constructivist exercises. That is, games are sometimes detected only by interpretation; sometimes games are made. Fourth, that many human activities are more, or less, conventional, and may be more, or less, rule-bound. In the ongoing discussion of the text-game analogy, it becomes increasingly obvious that games are primarily rule-governed, while literature is primarily convention-laden. Chess is an example of an activity that is weighted toward rule and away from convention; the sonnet (as a combination of regulated form and conventional content) is weighted toward convention and away from rule. The presence of rules does not preclude the presence of convention, or vice versa.

Although the text-game analogy is not precise—and I do not wish to argue that it is—it is nonetheless useful to keep the gamefulness of postmodernist literature playfully in mind during the discussion that follows.

The strong theory governing games and rules and literature rules out literature as game, utterly dismissing the possibility; the weaker version, the one which I am arguing for here, is that, while postmodernist literature is not a game in the strong sense—that is, in the sense of having a set of rules that is laid out and explained explicitly from the first or in the sense of having winners and losers and clear, unambiguous instructions—it is gamelike and gameful in the weak sense—that is, in the sense that it plays (it is, in fact, play), but that it plays within limits and borders and rules.

While literature generally is not a game, postmodernist literature is game-like, or, to use the term that titles this book, gameful. It must be seen that way, or else the endless play that is postmodernism will not resolve itself into the postmodern novel.

CHAPTER FOUR

John Barth: Becoming Somebody

> ... or some other such piece of antiquity as will exert a benign influence on the mind by giving it a hankering for impossible journeys through the realms of time. (Proust, *The Remembrance of Things Past*)

1. Barth the Player

"Tales within tales within tales, told for the sake of their mere marvelousness" ("Some Reasons Why...," *The Friday Book* 9): that is the way John Barth describes the *Book of the Thousand Nights and a Night*, and other story cycles, in his collection of assorted short non-fiction writings entitled *The Friday Book*.[1] It is with a sense of true wonder that he recounts in *The Friday Book* his falling in love with the marvels of various Oriental story cycles while pushing library carts full of books through the halls of the Johns Hopkins University library system.

Barth's writing, too, aspires to the lofty status of "marvelousness." Barth's novels are, indeed, often loving forays into the "marvelous" worlds of myth and old story: *Chimera*, for instance, retells the stories of Greek mythical heroes Bellerophon and Perseus (among others); an obscure but "real-life" Maryland poet of satirical verse named Ebenezer Cooke and his poetry are the subjects of *The Sot-Weed Factor*; Twain's Huck Finn, Cervantes's Don Quixote, Homer's Nausicaa and Odysseus, and a host of other literary characters, ancient and modern, make appearances in both *Sabbatical* and *The Tidewater Tales*; and, of course, Scheherazade appears more often than any other character in Barth's writing, fiction and non-fiction alike.[2]

Barth has claimed in many places to have a special fondness for Scheherazade, who, as an image of the storyteller, he says, is "the aptest, sweetest, hauntingest, hopefullest I know"; she is, moreover, "my condition and my hope, musewise" ("Muse, Spare Me," *The Friday Book* 56, 57). The author character in *Chimera* (a Genie in the book, and a thinly disguised Barth) claims that the first of three wishes he would make before death would be to meet Scheherazade in person; the second would be to make love last (this is important in that story and, perhaps, in "real life," but is not, here, in this book); the third would be to add "some artful trinket or two, however small, to the general treasury of civilized delights, to which no keys [are] needed beyond goodwill, attention, and a moderately cultivated sensibility" (*Chimera* 17). Those are relatively modest aims, considering that the author in *Chimera* has just met Scheherazade. One wish granted; good luck (I suppose) the arbiter in the second; and posterity the judge in the third.

On one hand, then, Barth appears to be a rather naive storyteller, fascinated by just story for story's sake, interested above all else in creating a sense of wonder and beauty with his fiction. In fact, he describes himself in terms suited to a straightforward (perhaps naive) realist, telling a story: "Once upon a time I told tales straight out, alternating summary and dramatization, developing characters and relationships, laying on bright detail and rhetorical flourish, et cetera" ("Anonymiad," *Lost in the Funhouse* 177). "Apt, sweet, haunting, and hopeful" are all descriptors of Scheherazade that suggest the notion of the storyteller as an "innocent"; and Scheherazade is, after all, "Islam's ... most celebrated virgin" (Tobin 162).

On the other hand, Scheherazade is also something else: she is also the consummate professional author and storyteller, both concerned and canny about her audience and his reactions to her tales. Scheherazade appeals to Barth as a compelling image of the storyteller, telling stories into the night, telling stories to stay alive—in a literal, and not merely a figurative sense.[3]

It is on this other hand, the one that holds the characteristics with which Barth is usually identified, that most of the stories from *Lost in the Funhouse* belong. Those stories are pointedly not straightforward or naive or "realistic." They include "Frame-Tale," a moebius strip the reader is to construct that reads "Once upon a time / There was a story that began," and so on; "Night-Sea Journey," a story about the purpose of life, narrated by a sperm; "Title," a story about titles that declares itself strangely placed (rather like one of Sterne's Dedications from *Tristram Shandy*) "past the middle, nearer three-quarters done, waiting for the end" (which is where the story appears in the book) and includes the much-quoted "Oh God comma I abhor self-consciousness" (105, 113); and "Menelaiad," a story made up of seven nested stories-within-stories, whose climax reads:

> " ' " ' " ' " " ' " ' " ' " (158)[4]

Barth's play, if sometimes rather self-indulgent, is purposeful. Among its more important purposes is to show that Barth is not merely wide-eyed at the wonder of literature; that he is not innocent or naive (that is, he is neither "Islam's most celebrated virgin," nor the virgin poet laureate of Maryland), but always (ironically) experienced and knowing; and to acknowledge that neither is his audience any of the above. Umberto Eco's reflections on postmodernism make this contrast between innocence and irony a central feature of postmodernism:

> The postmodern reply to the modern consists of recognizing that the past, since it cannot really be destroyed, because its destruction leads to silence, must be revisited: but with irony, not innocently. I think of the postmodern attitude as that of a man who loves a very cultivated woman and knows he cannot say to her, "I love you madly," because he knows that she knows (and that she knows that he knows) that these words have already been written by Barbara Cartland. Still, there is a solution. He can say, "As Barbara Cartland would put it, I love you madly." At this point, having avoided false innocence, having said clearly that it is no longer possible to speak innocently, he will nevertheless have said what he wanted to say to the woman: that he loves her, but that he loves her in an age of lost innocence. If the woman goes along with this, she will have received a declaration of love all the same. Neither of the two speakers will feel innocent, both will have accepted the challenge of the past, of the already said, which cannot be eliminated; both will consciously and with pleasure play the game of irony.... But both will have succeeded, once again, in speaking of love. (*Postscript* 67–68)

And Barth's writing, for all its aspirations to "marvelousness," is almost always aware of itself as problematic and problematized; it is always ironic in Eco's sense of that term. It is, in Barth's terms, rarely "prosaic"; it is rarely "'straightforward' or 'naive' as opposed to an oblique or self-reflexive manner of storytelling" ("The Role of the Prosaic in Fiction," *The Friday Book* 82).

Barth himself has never been (publicly, at least) naive or idealistic about writing; he claims, for instance, in "Some Reasons Why..." that one very strong reason he began writing in the first place was precisely "to get out of here [Dorchester County, Maryland] and become a distinguished something-or-other" (4). Elsewhere he claims to be "no amateur"—"I know my trade" (*Lost in the Funhouse* 177); and still elsewhere, he is interested not merely in the wonder of stories, but in the technical stuff that makes story work: "the original springs of narrative" fascinate him (*Chimera* 10).

In other words, for all the marvelousness and the playful self-referential self-indulgence, Barth is, after all, a professional who appears to have taken up writing from the first as a deliberate professional choice. Accordingly, perhaps, he has thought carefully about his craft and his profession, and takes those reflections seriously enough to not only write challenging and creative fiction, but to teach creative writing as well. And his professionalism has made him into not only a teacher of literature and a prolific writer of fiction, but an essayist who is respected for his reflections on his profession as well. Two of his so-called "Friday pieces" ("The Literature of Exhaustion" and "The Literature of Replenishment"), for instance, are engaging reflections on what postmodernism is, if it is anything; on what it means, if it means anything; and on who, exactly, is a postmodernist, if anyone is. Barth describes postmodernism succinctly as crossing generic, spatial, and temporal boundaries: "... we may regard ourselves as being not irrevocably cut off from the nineteenth century and its predecessors by the accomplishment of our artistic parents and grandparents in the twentieth, but rather as free to come to terms with both realism and antirealism, linearity and non-linearity, continuity and discontinuity. If the term 'postmodern' describes anything worthwhile, it describes this freedom, successfully exercised" ("The Spirit of Place," *The Friday Book* 128–29). That description of postmodernism is also a quite accurate description of *The Last Voyage of Somebody the Sailor*, which is a novelistic working out of the "freedom" that Barth sees as a strength of the postmodern.

For all his yearning after marvelousness, then, Barth sees himself also as a working professional, his telling as professional work: "telling stories," he says, "is my passion and vocation" ("Speaking of *LETTERS*," *The Friday Book* 176). At times, the metaphor he uses for this passion and vocation is that of a jazz musician: "At heart I'm an arranger still, whose chiefest literary pleasure is to take a received melody—an old narrative poem, a classical myth, a shopworn literary convention, a shard of my experience, a

New York Times Book Review series—and, improvising like a jazzman within its constraints, reorchestrate it to present purpose" ("Some Reasons Why...," *The Friday Book* 7). At other times, the metaphor is of a seasoned and capable lover: "Making love and telling stories both take more than good technique.... Heartfelt ineptitude has its appeal ... so does heartless skill. But what you want is passionate virtuosity" (*Chimera* 23–24).[5]

Those two metaphors are at the heart of *The Last Voyage of Somebody the Sailor*: Barth the jazzman has taken "a received melody," one of the better known Eastern story cycles, Scheherazade's Sindbad stories, and Barth the lover has married those stories to perhaps the most familiar form of Western narrative, the coming-of-age autobiographical novel. Brian Edwards notes that "it is the task of *Letters* to recycle and invigorate the 'out-of-date'" (169). That task is something that Barth has set himself not only in *LETTERS*, which is the immediate concern of Edwards, but in many other works as well. Barth recycles in the way Eco uses irony: he needs to tell a story, but he needs to tell it in an age that is no longer innocent, and therefore to tell it in a way that is no longer innocent; he thus makes reminding an audience that it is no longer innocent part of what is told. In *The Last Voyage of Somebody the Sailor*, Barth uses Scheherazade's stories as a framework upon which to hang his own material, thereby creating a nested story cycle, but creating it in a way that acknowledges that story cycles like *The Last Voyage of Somebody the Sailor* are ancient, not modern: Barth's is a postmodern story cycle; he borrows what is needed, adds what is lacking, crosses borders and boundaries when he needs to do so.

In the same way that Barth is fond of the image of himself as a word-oriented jazzman, he is equally fond of the image of the story-teller as a seductive lover, with narrative itself acting as a medium of exchange—almost a kind of sexual intercourse—between writer or teller and reader or listener: this, perhaps, is the "passionate" part of Barth's "passionate virtuosity." "A story first heard," he writes, "is a virgin bride, who so takes us with her freshness that we care nothing for her style. A good tale retold is a beloved wife or long-prized lover, whose art we relish because no novelty distracts us. Tell, this beggar begs you: from start to finish, tell, and while these worthies relish your grace notes and flourishes, my worthless ears will hang on the tale's mere melody. Do tell!" (*The Last Voyage of Somebody the Sailor* 21)

The twin figures of the jazzman and the lover are at the heart of what *The Last Voyage of Somebody the Sailor* is; they are also emblematic of what makes postmodernism gamelike. It is these two modes, these two ways of writing, and the boundaries and boundary crossings between them and among the worlds that they represent, that *The Last Voyage of Somebody the Sailor* is fundamentally about. I have already suggested that *The Last Voyage of Somebody the Sailor* is obsessed with the crossing of boundaries: tenuous ones, like the one between "serious" and "playful," for instance; or more palpable ones, like those between Eastern and Western; or numerous others, like the boundaries between reality and the imaginary, or realism and the fantastic, or life and autobiography, or journalism or fiction, or between men and women. *The Last Voyage of Somebody the Sailor* is about approaching and crossing boundaries, about the tenuous nature of the distinction between worlds in "life" and worlds in "literature." *The Last Voyage of Somebody the Sailor* implicitly refers to works like *Don Quixote*, whose hero cannot distinguish between giants lifted from the pages of his romances, and windmills; or to *Huckleberry Finn*, wherein Tom Sawyer is unable to discern differences between "Spaniards and A-rabs, camels and elephants" and a gathering of children that, according to Huck, "had all the marks of a Sunday School." In Barth's boundary-crossing postmodernist fiction, that foundered steamship of romance, the "Walter Scott" from the middle of *Huckleberry Finn*, would have been refurbished, refitted with oriental sails, and then used to sail the voyage of Poe's "Sinbad."[6]

The crossing of boundaries in *The Last Voyage of Somebody the Sailor* is truly an obsession throughout: when Simon William Behler is a child, for instance, he rocks himself to sleep, thereby moving away from his world, and toward another: "My closest approaches thus far to anything resembling the boundary of those constraints [that is, of time, place, identity].... had been ... when ... I was able in a certain combination of drowsiness and less-than-total darkness to rock myself just beyond all usual and normal sensory cues into a charged suspension, vertiginous, electrically humming ... and I myself ... the very lens of the cosmos" (*The Last Voyage of Somebody the Sailor* 107). Even earlier in the novel than the place where Simon's rocking appears, however, the ordeal of birth is described as a literal movement from one world into another. Simon and his twin sister "Bijou" are frightened but eager about that passage: "We know it's risky, this shift from one world to another; we're edgy but eager" (27). And this novel that begins before birth ends with a second literal passage, this one into death. Bijou is now urging: "You went first, you always thought, but *I* was the one; *I* went first. I know the way. Follow me now: Two. One" (573). There are countless other images of Simon's fascination with boundaries and the limits of worlds: when he swims in the Chaptico as a boy, he "float[s] on [his] side, one eye above the surface, one below: different worlds" (42). When he experiences sex for the first time (that experience is itself seen as a kind of passage from boyhood into manhood), "[Daisy] drew my climax vertiginously out, right to that Boundary that I had fancied Home Plate might even carry us beyond" (130). When he wakes from sleep, it is "From near-dreams of near-drownings at the Boundary—Island Point on the Chaptico, Soper's Hole, Adam's Peak—a piping piped him briskly up, to and through the surface" (282). When he reads fantasy tales at Daisy's mother's urging, he notes, "My recent reading in Mrs. Moore's *Arabian Nights* had made me chafe not only at being ineluctably I and here and now but likewise at the iron constraints of nature itself, which made it quite certain that no fish would really ever talk and no genie appear from a bottle, nor would Daisy and I be magically transported from Dorset County to Samarkand or Serendib" (107). Ironically, when "Baylor" finally achieves a passage from his own world to another, his nickname becomes "Still-Stranded" (precisely because he is in a world not his own, and wants to re-negotiate a boundary he had been trying all his life to cross in the first place). These images of boundary crossings appear at consistent places in the novel: they appear at the literal passages of birth and death, prompted by celebrations of Simon's birth (his various birthday celebrations become his voyages; they gradually become literal voyages—birthday cruises), and occasions of his death (he drowns on his cruise with Julia Moore, for instance). They also appear during particular activities: moments of play, during sex, while dreaming, and when reading. It is at those moments, that is, of Simon's greatest involvement with the imaginary, when he is least involved with the physical constraints of his world and himself, that boundaries can be approached and sometimes crossed.

I have been suggesting that this is a novel that is in part about boundary crossings and about the Protean nature of our many definitions of the world or worlds in which we live; it is a postmodernist novel in its obsession with exploring those boundaries, and in its suggestion that all worlds are as elastic as Simon's: "we carry ourselves inside our heads—our time and place, who and what we are" (107). The method Barth has used to construct his various worlds is itself a postmodernist one, and one that is ineluctably tied to the way we (literally) see the world.[7] Barth notes:

> Even the detail that Scheherazade's stories are drawn from the literal and legendary foretime, I find arresting. It reminds me that the eschewing of contemporaneous, "original" materials is a basic literary notion, by comparison to which its use is but an occasional anomaly and fad of the last couple of centuries. Not only classical epic and tragedy, and Elizabethan and neoclassical drama, but virtually all folk and heroic narrative, both Eastern and Western, follows Horace's advice:
>
> ... safer shall the bard his pen employ
> With yore, to dramatize the tale of Troy,
> Than, venturing trackless regions to explore,
> Delineate characters untouched before.
>
> ("Muse, Spare Me," *The Friday Book* 58)

The dilemma for Barth, then, is the problem of how to reconcile his love of the methods and stories of antiquity with modern demands for originality. It was also Umberto Eco's problem:

> Is it possible to say "It was a beautiful morning at the end of November" without feeling like Snoopy? But what if I had Snoopy say it? If, that is, "It was a beautiful morning..." were said by someone capable of saying it, because in his day it was still possible, still not shopworn? (Eco, *Postscript* 18)

Umberto Eco's solution in *The Name of the Rose* is to write a murder mystery set in a medieval monastery and narrated by a medieval monk; that narrative is then filtered through several layers of textuality and intertextuality. The result is Eco's novel, a postmodernist medieval romance and murder mystery.

Barth's solution is similar: his novel is a late-twentieth-century piece of autobiographical fiction written as the memoirs of Baylor, a "new journalist," and written as a story cycle that is then traded, voyage for voyage, with Sindbad's "original" accounts. *The Last Voyage of Somebody the Sailor* is a postmodernist medieval Arabic American novel that incorporates multiple worlds and multiple kinds of storytelling into itself.

The critical apparatus that has grown up around the work of John Barth is enormous. Much of that critical interest centres on Barth's play and playfulness, on his "rich, protracted playful works" (Fogel, *Understanding* 214). Brian Edwards, for instance, calls Barth a "recycler" in connection with his novel *LETTERS* (183); Benzi Zhang writes of Barth as a paradigm case of playful self-reflexiveness—an outstanding practitioner of "Chinese-box fiction"; Patricia Tobin eloquently discusses Barth's "fanciful and fortuitous" (165) fictions; or the "magical" (167) goings on in novels like *The Last Voyage of Somebody the Sailor*.

As will have been clear in my earlier discussion, "boundary" is something of a magical word in this novel. The particular boundaries that I wish to point out are not really the magical ones, but the ones that specifically set the limits and the borders of this novel. They are the intertextual references which abound here; they are the works that define what, exactly, this novel is and where it is situated in the world of literature, that both expand and limit this novel, that situate it in the world of writing and telling that makes up "The Sea of Stories."

Julia Kristeva defines intertextuality in this way: "in the space of a given text, several utterances, taken from other texts, intersect and neutralize one another" (36). Kristeva's notion of intertextuality is a common property to all texts. The implication is that, whenever we read a novel, we are also always reading The Novel—all novels, in a manner of speaking. The references that so abound in *The Last Voyage of Somebody the Sailor* are the results of specific reference; they are more like the kind of quotation that Eco discusses in his description of postmodernist irony than Kristeva's intertextuality, which has to do with the nature of the Novel. What Barth is up to is intertextual, but it is an intertextuality that defines this particular novel, while at the same time establishing it as Novel.

I shall not pretend to have exhausted all the possibilities and references that abound in *The Last Voyage of Somebody the Sailor* in this discussion; what I aspire to in this discussion is to portray a sense of the range of references that are here, the importance of literary reference to this novel, and the approximate location of this novel in the world of Novel, the Sea of Story.

2. Barth the Gamesmaster

A. *The Book of the Thousand Nights and One Night*

The first of the major intertexts of *The Last Voyage of Somebody the Sailor* is, of course, *The Book of the Thousand Nights and One Night*.[8] The primary frame story of Barth's novel is the story of the death of Scheherazade; she agrees to tell to "the Destroyer of Delights" (the character who traditionally brings to an end Eastern narratives like *The Thousand Nights and One Night*) a story in exchange for his visitation; she wishes to die, to cross "the Boundary," to finish her life's "story." The primary frame story of *The Book of the Thousand Nights and One Night*, on the other hand, is the story of Scheherazade telling stories to *avoid* death, to *avoid* that Boundary. Barth has claimed in several places that "it was never Scheherazade's stories that seduced and beguiled me, but their teller and the extraordinary circumstances of their telling: in other words, the character and situation of Scheherazade, and the narrative convention of the framing story" ("Tales within Tales within Tales," *The Friday Book* 220). Barth's account of the frame story of Scheherazade makes literal the closure that is already partly present at the end of the original frame story:

> King Shahryar and Queen Shahrazad, King Shahzaman and Queen Dunyazad, and Shahrazad's three small sons, lived year after year in all delight, knowing days each more admirable than the last and nights whiter than days, until they were visited by the Separator of friends, the Destroyer, the Builder of tombs, the inexorable, the Inevitable. (4: 536)

Barth's novel's frame is an account of the story of the story with which Scheherazade finally seduces the Destroyer of Delights into taking her beyond "the Boundary."

Barth's use of frames from *The Thousand Nights and a Night* in *The Last Voyage of Somebody the Sailor* does not end with Scheherazade, however. The character of "Somebody" in *The Last Voyage of Somebody the Sailor* is himself also borrowed; he is also a frame. He is based on a rather strange minor character in Scheherazade's original Sindbad stories:

> It is related ... that there lived in Baghdad, during the reign of Khalifah Harun al-Rashid, a poor man called Sindbad the Porter, who earned his living by carrying loads upon his head.... Stirred to his depth by such thoughts, the porter made up ... stanzas and sang them at the top of his voice.... (2: 176)

First, the minor character named Sindbad the Porter[9] who appears in Sindbad the Sailor's story is at least by name a twin (and all readers of Barth know how fond he is of twins and twinning[10]).

Second, Sindbad the Porter is himself part of a framing device situated well inside the framing device of the story of the story of Scheherazade, her travails, and her tale sequences: a tale within a tale (and so on), and therefore a perfect candidate for usage by Barth, who likes Scheherazade most, he confesses, for her use of "the narrative convention of the framing story."

Third, Sindbad the Porter is the person to whom Sindbad the Sailor tells his stories: "I have accomplished seven extraordinary voyages, and the narrative of each one is enough to stupefy listeners with an excess of marvel" (2: 178). The stories are told with the view to demonstrating a moral, that the life of adventure that Sindbad the Sailor has lived should not necessarily be envied, in comparison to the "more tranquil life" of Sindbad the Porter: "Destiny" favours whomever she pleases (234), concludes Sindbad the Sailor; adventures are very hard work. Sindbad the Porter is the audience in Scheherazade's story; he is the one for whom the "moral" is voiced. An audience turned inside out becomes an author, the other half of a set of twins, a mirror image; Sindbad the Porter is the obvious structural choice for the narrator of Barth's parallel story cycle.

Fourth, we know nothing of Sindbad the Porter's life until he comes under the influence of his more illustrious namesake: he comes to the door of Sindbad's house apparently by chance, and, we are told, never does leave: the two Sindbads "lived together in perfect friendship and joy until they were visited by That which breathes upon delight, which snaps the links of friendship, which destroys palaces and raises tombs where once they stood: by bitter Death" (234). Sindbad the Porter must have a life, and therefore, in this world of storytelling, a story to tell, before he wanders into the household of Sindbad the Sailor: for an author with Barth's predilections, this is a gap that needs filling. As it turns out, in Barth's account, Sindbad the Porter ("Somebody") has an even more adventurous life than Sindbad the Sailor has had: he voyages through time and between worlds, not just through space and among "real" islands.

Fifth, is the happy circumstance that Sindbad the Porter makes up "stanzas." He is a poet. We meet him as he stands outside Sindbad the Sailor's house singing his inventions: Sindbad the Sailor claims that his stories are taken from his life; Sindbad the Porter, on the other hand, obviously deals in imagination. Of the two, Sindbad the Porter more closely matches the figure of the teller of imaginative tales, the figure of the writer of fiction. Sindbad the Sailor's tales may "stupefy ... with an excess of marvel," but they claim nonetheless to be true. They claim to be autobiography. Sindbad the Sailor is an autobiographer, a historian; Sindbad the Porter, however, is an artist strictly of the imagination, a maker of fiction.

B. *"The Thousand-and-Second Tale of Scheherazade"*

The second, and perhaps even more pertinent intertext for *The Last Voyage of Somebody the Sailor* than the "original" Sindbad stories, is Edgar Allan Poe's "The Thousand-and-Second Tale of Scheherazade." Poe has appeared elsewhere in Barth's work—he is important, for instance, in *The Tidewater Tales*—and Barth's debt in this case likewise is striking. First, although Barth may owe his abiding preoccupation with framing stories to Scheherazade, he appears to owe the particular structure of the framing story in *The Last Voyage of Somebody the Sailor* to Poe. Poe's story begins this way:

> Having had occasion, lately, in the course of some Oriental investigations, to consult the "Tellmenow Isitsoörnot," a work which (like the "Zohar" of Simeon Jochaides) is scarcely known at all, even in Europe; and which has never been quoted, to my knowledge, by any American—if we except, perhaps, the author of the "Curiosities of American Literature";—having had occasion, I say, to turn over the pages of the first-mentioned very remarkable work, I was not a little astonished to discover that the literary world has hitherto been strangely in error respecting the fate of the vizier's daughter, Scheherazade, as that fate is depicted in the "Arabian Nights"; and that the *denouement* there given, if not altogether inaccurate, as far as it goes, is at least to blame in not having gone very much farther. (280)

Poe promises to tell here the "true" story of the death of Scheherazade; his story is a corrective to the original "Arabian Nights." Barth's novel, likewise, begins:

> The machinery's rusty, I acknowledge to my half-expected guest, but it hasn't seized up altogether. "I could tell you one about the death of Scheherazade...."
>
> "I've heard it."
>
> "Not this version."
>
> My visitor crosses her nyloned legs with stern self-assurance and smooths her skirt-pleats. I was rather expecting an elder man, cordially disposed but still skeptical.... "I've heard them all."
>
> "Those were King Shahryar's very words," I declare to her. "In fact, his *last* words." (*The Last Voyage of Somebody the Sailor* 3)

The general conceit of the two stories is thus in place: each promises to correct the original, flawed story of Scheherazade's death.

Barth has borrowed not only the general conceit of his novel from Poe, however, but the specific structure of his novel as well. Poe's stories have always dealt with the tenuousness of the boundaries between real and unreal, truth and the fantastic, natural and supernatural: his tales propose that murder, horror, entire other worlds, exist beside our familiar, everyday one: "The Black Cat" is an excellent example. It is a story wherein an apparently ordinary man decides that a black cat is tormenting him; in a fit of rage, he hangs the cat. One event follows another, until, in trying to kill a second cat, the man kills his wife. That second cat (apparently the reincarnation of the first) gains its revenge by attracting the police to the place where the woman's body is walled up in the basement: Poe's story demonstrates with startling clarity how ordinary life can transform into horror, sanity into madness. Poe's "The Black Cat" makes its own preoccupation with distinctions (boundaries?) between the "real" and the "unreal" explicit. It begins:

> For the most wild yet most homely narrative which I am about to pen, I neither expect nor solicit belief. Mad indeed would I be to expect it, in a case where my very senses reject their own evidence. Yet, mad I am not—and very surely I do not dream. But to-morrow I die, and today I would unburden my soul. (550)

When Poe's Scheherazade decides to "redeem the land from the depopulating tax upon its beauty" (281) by telling stories to her husband, she begins, as it happens, with one of Poe's own tales, "about ... a black cat, I think" (282). In Poe's account, Scheherazade's story is thus announced as a version of Poe's own story, which is itself a story about crossing boundaries between the real and the marvelous. In turn, this is, of course, exactly what Barth's novel is about.

Poe goes on to explain that Scheherazade, "on the thousand-and-second night," tells "the full conclusion of Sinbad [sic] the sailor. This person went through numerous other and more interesting adventures than those which I related; but the truth is, I felt sleepy on the particular night of their narration, and so was seduced into cutting them short" (283). Poe offers the "true" story of the death of Scheherazade, and will tell that story by adding one further tale to the Sindbad story cycle. That is, of course, exactly what Barth proposes to do in *The Last Voyage of Somebody the Sailor*.

Even in the earliest "voyages" by Somebody, there are echoes, faint but indisputable, and extremely naive, of the exotic oriental world to come:

> We boys attended Sunday school.... I myself took more interest in the exotic but vaguely depressing world of the illustrated lessons—camels and palm trees, figs and dates and anointing-oil, sandals, robes, unbarbered hair and beards—than in the strange agonies and paradoxical imperatives of the lessons themselves, presided over by cheerful matrons of the neighborhood: Whoso loseth his life shall find it, and the rest. Let the lower lights be burning, we sang together at the end:
>
> Send a gleam across the wave,
> Some poor sinking, suff'ring seaman
> You may rescue, you may save. (33)

Poe's story likewise draws important distinctions between his Western world and the Eastern one of Islam. These differences are most clearly summarized in what stories the king chooses to believe:

> "It was just after this adventure that we encountered a continent of immense extent and prodigious solidity, but which, nevertheless, was supported entirely upon the back of a sky-blue cow that had no fewer than four hundred horns." [Poe inserts here a footnote reference to *Sale's Koran*]

> "*That*, now, I believe," said the king, "because I have read something of the kind before, in a book." (297)

The king is able to believe stories like the one about the blue cow "because [he had] read something of the kind before," notably in the *Koran*. But he refuses to believe Sindbad's stories of Western marvels.

Sindbad's tale of the steamship on which he travels about the world, for instance, is only accepted by the king when the boat is described in fantastic terms as a fearsome beast: it is like a roc, big enough to ride upon. Poe includes, on the other hand, a story about a printing press that "had no trouble in writing out twenty thousand copies of the Koran in an hour; and this with so exquisite a precision, that in all the copies there should not be found one to vary from another by the breadth of the finest hair" (299). This the king pronounces "Ridiculous." Scheherazade's account of Sindbad's account of the Mammoth Caves of Kentucky, one of the natural wonders of the United States, and a prized discovery of western natural science (itself in its heyday in the nineteenth century), is met with similar incredulity: "'We reached a country where there was a cave that ran to the distance of thirty or forty miles within the bowels of the earth.... and in among the streets of towers and pyramids and temples, there flowed immense rivers as black as ebony, and swarming with fish that had no eyes.' 'Hum!' said the king" (291). The king is equally skeptical about the petrified forests of Texas, or coral islands "built in the middle of the sea by a colony of little things like caterpillars" (290). He will only believe in, or accept stories about, things with which he is already familiar: the blue cow from the *Koran* and the steamship, if it is described not as a technological marvel, but as a great and fearsome beast, are acceptable to him; but he refuses to believe the story about the caves of Kentucky, the coral islands, the petrified forests of Texas. In short, he refuses to believe any of the stories that Poe considers fact and which he supports with footnotes: the king values imagination, storytelling, and religion over science, technology, and documentation.

The attitude of Poe's king is echoed strongly—and tellingly—in Barth's novel as well. Somebody's audience is skeptical of his stories, examples of traditional Western realism, because they are not the kind of stories that that audience is used to:

> The high ground of traditional realism, brothers, is where I stand! Give me familiar, substantial stuff: rocs and rhinoceri, ifrits and genies and flying carpets, such as we all drank in with our mothers' milk and shall

> drink—Inshallah!—till our final swallow. Let no outlander imagine that such crazed fabrications as machines that mark the hour or roll themselves down the road will ever take the place of our homely Islamic realism, the very capital of narrative—from which, if I may say so, all interest is generated. (136)

The "high ground of traditional realism" in the East, however, is not the same as the "high ground of traditional realism" in the West.

Barth has claimed that autobiographical fiction, which both Somebody's stories and Baylor's "new journalism," for instance, resemble, is "a genre I have no use for" ("Aspiration, Inspiration, Respiration, Expiration," *The Friday Book* 98); Fenn, accordingly, proclaims in *Sabbatical* that "I won't have our story be unadulterated realism. Reality is wonderful; reality is dreadful; reality is what it is. But realism is a fucking bore" (136). In *The Last Voyage of Somebody the Sailor*, "realism" becomes in effect and fiction to which the audience has become accustomed.[11] No one believes that rocs really exist, but rocs are treated as though they might, in stories and by storytellers, because that is what the audience wants:

> Unruffled, Somebody said, "Grown men may swallow what a child would choke on: whales the size of islands, with landscapes on their backs! Rocs and rhinoceri! Shores strewn with diamonds!"
>
> Sindbad smiled. "Because their appetite was whetted with greed, friend, and the marvels washed down with wine."
>
> "As they will be tonight," said Somebody, "and tomorrow and the evening after. But in my stories there are no genies or one-eyed giants to be swallowed whole. Nature's laws are not transgressed."
>
> "Only the laws of human nature," Sindbad replied, "which bid hungry men eat when food is in their hands, as Yasmin was in the hands of We-Know-Whom. Besides, I brought real diamonds home...." (71)

The king in Poe's story is criticized because he is unwilling to listen to any story that does not fit his criteria for truth and proper storytelling; Poe comments that he is fittingly "[deprived] ... of many inconceivable adventures" (302) because he is absolutely insistent that the world of the stories match his own impression of the world of life and literature. He is, in short, very much like the critics in Barth's novel, who think Somebody's stories are crude and worthless.

If the attitude of Somebody's audience resembles the attitude of Poe's king, it resembles also the attitude of Barth's many critics, who have often derided not only Barth but postmodernism more generally. Charles Newman's is a typical, if eloquent, antagonistic response to postmodernism: "Never before in history has so much of a literary heritage been available to so broadly based an audience, yet never before has a culture drawn such apparently meagre sustenance from its own literary activity" (15). Newman wants "sustenance"; Barth wants to provide "mere marvelousness." No wonder the two are at cross purposes.

The story which leads to the end of Scheherazade's life in Poe's story is the most distinctly Western one in her entire collection; Scheherazade includes a story about the bustle, and is "throttled" because of it:

> "One of the evil genii ... has put it into the heads of ... accomplished ladies that the thing which we describe as personal beauty consists altogether in the protuberance of the region which lies not very far below the small of the back.... Bolsters being cheap in that country, the days have long gone by since it was possible to distinguish a woman from a dromedary—"
>
> "Stop!" said the king—"I can't stand that, and I won't. You have already given me a dreadful headache with your lies.... And then that dromedary touch—do you take me for a fool? Upon the whole, you might as well get up and be throttled." (302)

In Poe's account, it is tension between East and West, symbolized most strongly in fashions in female clothing, that leads to Scheherazade's "throttling."

This tension informs not only Poe's story, but Barth's as well. Barth's novel, after all, was published contemporaneously with the first Gulf War; it follows closely not only that war, but the earlier scandals surrounding the Ayatollah Khomeni, the Shah of Iran, and various other tensions between the United States and Middle Eastern Islamic cultures.[12] It thus becomes extremely important that Behler's daughter Juliette lives with "an expatriate Omani doing graduate work in oceanography" (194), and eventually marries him; Barth's Juliette is a knowing nod to *Romeo and Juliet*, with its feuding parents and suggestion that boundaries like the ones that cause feuds should be crossed.

Finally, the general structure of Poe's story echoes a significant feature of the Burton translation of *The Thousand Nights and a Night*. That structure includes the exhaustive footnoted references and other efforts to normalize the *Nights* for Western audiences by providing documented Western information.[13] Somebody knows that audiences can be fickle and demanding, although at least a minimal audience is necessary for telling: "Every teller ... tells at his audience's pleasure. I would be most pleased if my story pleased all who heard it. Failing that, I shall be satisfied if it satisfies my host and any other whose heart is open to it and its teller. Failing that, I'll tell on all the same while I have one listener for whom I care" (137). However, he also knows what Barth suggests elsewhere: "Just as the accounts of Hakluyt's voyagers may be more fascinating than the places they voyaged to, the Burtonesque notes and appendices ... to Penzer's edition of Somadeva ... are frequently more engaging than the texts they illuminate.... One's conviction is affirmed that it would be a more splendid destiny to have cooked up *Burton's* version of *The Thousand and One Nights*—footnotes, Terminal Essay, and all—than to have written the original" ("*The Ocean of Story*," *The Friday Book* 86–87).

The movement of Somebody's stories is from straightforward Western autobiographical realism toward the fantastic world of Sindbad's adventures: the two (worlds, tellers, voyages, stories) meet in the respective fifth and sixth voyages of the respective Sindbads. Barth's novel argues (if it can be said to "argue" anything) that although Western literature may have drawn inspiration from stories like *The Thousand Nights and a Night*, modifying those stories for its own uses, real maturation involves immature realism growing toward the more venerable fantastic: "a story with nothing fantastic in it lacks something essential.... But what we're after is the Truly Irreal; the Literally Marvelous.... The literally marvelous is what we want, with a healthy dose of realism to keep it ballasted.... Realism is the ballast. I like that" (*Sabbatical* 135–36).

C. *Native Son*

"With a promptness more Brit than Carib, at 0800 almost sharp the native captain of *Native Son* sounded the ship's whistle" (*The Last Voyage of Somebody the Sailor* 193). *Native Son* in *The Last Voyage of Somebody the Sailor* is a ferry that runs "from West End, Tortola, in the British Virgins, for Charlotte Amalie, St. Thomas, in the U.S. Virgins" (192). *Native Son* is yet another instance of boundary crossing; this one is a multiple crossing. The setting in Barth's novel is the Virgin Islands, a combination of Caribbean and British cultures, on one hand, and Caribbean and American cultures, on the other. The "native captain," we are told, is himself a combination, being "native," yet possessing British "promptness." *Native Son* in Barth's novel is a ferry, a vehicle designed to cross boundaries, both literary and actual: the ferry takes "Baylor" and his family to Charlotte Amalie, where one can take a wrong turn and end up in a medieval marketplace.

Native Son is also a novel about crossing boundaries—racial, social, and economic ones. It is the story of Bigger and his inability to cope in a world full of difference and discrimination: the rich take advantage of the poor; Whites take advantage of Blacks; men take advantage of women. Bigger crosses a boundary of acceptability into rape and murder; Wright's novel recounts the elimination of a boundary crosser by the enforcers of the boundaries.

The Introduction to Richard Wright's *Native Son* begins:

> I am not so pretentious as to imagine that it is possible for me to account completely for my own book, *Native Son*. But I am going to try to account for as much of it as I can, the sources of it, the material that went into it, and my own years-long attitude toward that material.
>
> In a fundamental sense, an imaginative novel represents the merging of two extremes; it is an intensely intimate expression on the part of a consciousness couched in terms of the most objective and commonly known events. It is at once something private and public by its very nature and texture. (vii)

Although Wright's novel plays a relatively small part in Barth's *The Last Voyage of Somebody the Sailor*, Wright's introduction not only captures quite precisely part of what Barth's novel is about, but also articulates the task that is faced by anyone attempting to write about Barth's novel. *The Last Voyage of Somebody the Sailor* is a series of extremes: on one hand, it claims the status of autobiographical fiction, being the life story of Simon William Behler who, following his father's instructions, decides to make "Somebody" of himself: that task is accomplished in a literal way, as it turns out, as well as the figurative way Mr. Behler means. On the other hand, it is a fantastical re-telling of one of the more familiar of Scheherazade's tale cycles. The novel claims explicitly, then, that it is both fact and fiction; it is both halves of precisely what Wright cites as a serious concern of every novel and every novelist. However, where Wright addresses directly the related problems of balancing objectivity with subjectivity, public versus private, life versus fiction in novels, Barth is oblique and playful, embedding the questions of origin and focus not in a formal introduction to his novel (as Wright has done), but using these tensions and juxtapositions as essential parts of his novel's structure. The questions that Wright raises as preamble and pretext for his novel, then, are already and always parts of fiction, for Barth. It is difficult to imagine a Barthian novel that could be anything other than obsessed with these questions, and with the dialectic that exists between and among them. The tensions between public and private discourse, between autobiography and story-telling, between life and fiction, make up much of what Barth's novel is about.

Native Son begins with an episode, after the famous opening "rat hunting" scene, that reappears, transformed, in *The Last Voyage of Somebody the Sailor*. Bigger and his friend see a plane sky-writing a commercial: "Use Speed Gasoline." That plane exists in a world far beyond the reality, or even of the dreams, of Bigger and Gus:

> "I *could* fly a plane if I had a chance," Bigger said.
> "If you wasn't black and if you had some money and if they'd let you go to that aviation school, you *could* fly a plane," Gus said. (20)

The differences between what Simon experiences and what Bigger experiences are so striking that they appear to have nothing at all in common. Simon lives in a white middle-class suburb: he is not black, he has some money, and "they" *will* let him go to aviation school. He not only gets a plane ride for his birthday, but he is capable of making decisions about his life. As it turns out, he chooses during his birthday flight not to become a professional pilot. The point is, the choice is his, and not up to others; his choice is based explicitly neither on race nor money, because implicitly his race and his money give him the luxury of choice:

> My boyish ambition to be a pilot, however, fell lightly away from me with the treetops and rooftops of Avondale.... commercial aviation ... was ... *an insufficient destiny*. (52–53)

Part of what bothers Bigger about his life is his lack of choice. There are arbitrary lines set up that he can't cross:

> We live here and they live there. We black and they white. They got things and we ain't. They do things and we can't. It's just like living in jail. Half the time I feel like I'm on the outside of the world peeping in through a knot-hole in the fence.... Why don't they let us fly planes and run ships.... (23)

While there is a formal similarity between Simon's yearning to cross boundaries and Bigger's frustration with boundaries, and though the limits of each world are discussed at times with similar terminology, Simon'' problems are considerably less urgent or visceral than Bigger's are: Simon's boundaries are metaphysical: "I rocked, and it was not long before I reached that state between two worlds where distances go strange and the familiar is no more. I was in black space, suspended, or as it were with one eye open above some surface, one below" (59). Bigger's boundaries are physical, and very different ones from Simon's:

> "You know where the white folks live?"
> "Yeah," Gus said, pointing eastward. "Over across the 'line'; over there on Cottage Grove Avenue." (24)

An assumption of Barth's novel, is that there are always choices: the very premises of the world that is called into question by Simon's yearning for magic is what has made that yearning possible.

Simon's unsatisfying "real" world is a fantasy world to Bigger, at least as unattainable to him as Somebody's world of "mechanical birds and bracelets that measure time" (60), or of "such crazed fabrications as machines that mark the hour or roll themselves down the road" (136) is to Sindbad and his guests. Bigger's reality is grim, bitter, violent; his dreams are as unreachable for him as the advertising planes that fly overhead.

Simon's reality, on the other hand, is troublesome to him because of its lack of imagination; he yearns for an escape to the world he has read of in "Mrs. Moore's *Arabian Nights*" (107). Bigger shows no indication that he can read; his world of the imagination consists not of "classic adventure literature," but of advertising slogans and propagandistic B-movies that he sneaks into that tell absurd melodramatic morality tales about rich white Americans and the ubiquitous Communist menace:

> "What's a Communist?"
> "A Communist is a red, ain't he?"
> "Yeah; but what's a red?"
> "Damn if I know. It's a race of folks who live in Russia, ain't it?"
>
> ....
>
> "Reds must don't like rich folks," Jack said.
> "They sure must don't," Bigger said. "Every time you hear about one, he's trying to kill somebody or tear things up."
>
> (Wright 34–35)

D. *Sailing Alone Around the World and Voyage of the Liberdade*

Near the end of *The Last Voyage of Somebody the Sailor*, Somebody and Yasmin decide to celebrate their wedding by enacting a scene out of a classical romantic adventure. They decide to spend their honeymoon building and sailing a boat: they will design and build their own small craft and sail Sindbad's seventh voyage, from Baghdad to Serendib, in it. The boat they build is described as "a scaled-up, Arabized version of Simon Behler's memory of Captain Joshua Slocum's famous *Spray*, in which that doughty New Englander in the 1890s made the first known solo circumnavigation of the globe" (*The Last Voyage of Somebody the Sailor* 564).[14] Slocum's voyage around the world ended in June of 1898; he took approximately three years to sail single-handedly around the world. The story of Joshua Slocum must have appealed to Barth on several counts. First, despite the reference in *The Last Voyage of Somebody the Sailor*, Slocum did not build the *Spray* from scratch in the way Yasmin and Simon build *Zahir 3* in *The Last Voyage of Somebody the Sailor*: Slocum was originally given "an antiquated sloop" (34), which was long out of salt water, covered with a piece of canvas and sitting in a field, almost beyond salvage. Slocum rebuilt the *Spray* based on what was left of her, using her structure and therefore her design, but adding all new materials of his own ("Better timber for a ship than pasture white oak never grew," he notes with satisfaction [35]). So the *Spray* itself is no virgin craft, but, like the stories in *The Last Voyage of Somebody the Sailor*, is refitted, recycled, and reused: the story of the *Spray* in this way greatly resembles the stories of all of Sindbad's voyages. This point is pressed home in *The Last Voyage of Somebody the Sailor*:

> What concerns me most, however, is their vessel. Not its low-tech, high-maintenance materials: sailors for centuries made do with wood and hemp, cotton and bronze; Josh Slocum's *Spray* was built of nothing more durable. But the design ... Simon William Behler knows himself to be no naval architect; a seasoned sailor but only a knowledgeable novice at marine design and construction, he has endeavored to meld his experience of modern materials and his memory of traditional designs with the very different conceptions and experience of medieval Arab shipwrights. (567)

As I have already suggested, Slocum's boat's structure and design were recycled: the materials were new, but the structure was in place before he started. In the same way that Barth claims that Scheherazade perfected but did not invent the design of the tale-within-a-tale, and delights himself in reusing that structure and design in his turn, Joshua Slocum has the good fortune to be able to add fresh, new materials onto an existing design: the *Spray* remains the *Spray*, in short. It does not matter that she has entirely new fittings, from keel to floor planks. This is the same fate that befalls Sindbad's stories in Barth's interpretation: the structure is there, the design is intact, but the expected amulets become wristwatches; airplanes take the place of flying carpets.

The second attraction of the Slocum story for Barth must surely have been the romantic narrative possibilities of a story about sailing alone, or with only a single companion. Barth's most recent novels have repeatedly featured sailing trips in small boats by pairs of lovers. And Slocum performed what is surely the ultimate small-craft feat: he sailed around the world in a boat he built himself. His is an example to emulate, or at least to borrow from.

The third attraction of the story is actually not part of the story per se: Joshua Slocum's adventures do not end with his successful "circumnavigation of the globe." In his Introduction to Slocum's book, Arthur Ransome notes:

> He went on sailing in the *Spray* after his return, taking his little vessel nearly every winter down to Grand Cayman in the West Indies. In 1909, at the age of sixty-five, he fitted out once more for a long voyage and in the autumn sailed from Bristol, R.I., for the Orinoco River. The *Spray* was never seen again.... On her projected voyage she would be crossing no less than five recognized steamer tracks ... she must have been cut down in the night. (27)

There are many examples of solo sailing in *The Last Voyage of Somebody the Sailor*. Simon's brother, Joe Behler Jr., for instance, survives a wreck and the resulting ordeal in an open boat on the Pacific; Sindbad's voyages virtually all have instances of survival after shipwrecks in open boats, in tubs, on spare timbers (Barth actually calls this "the classic Sindbad situation: once again derelict in Allah's ocean" [567]). The Joshua Slocum adventures most resemble, however, what must have been the various explanations of Behler's disappearances from one world or the other in *The Last Voyage of Somebody the Sailor*: "Baylor" disappears within sight of Tim Severin's rescue dinghy, only to reappear, stranded, in the world of Sindbad and Yasmin; and then he apparently drowns in a storm in "Allah's ocean," only to reappear on a shore in Maryland. Slocum, too, disappears without trace, at least in his "real" world; he resurfaces in Barth's novel.

The fourth significant point of comparison between *The Last Voyage of Somebody the Sailor* and Slocum's adventures is the relative importance of chronographs to navigation in each:

> The want of a chronometer for the voyage was all that now worried me. In our newfangled notions of navigation it is supposed that a mariner cannot find his way without one; and I had myself drifted into this way of thinking. My old chronometer, a good one, had been long in disuse. It would cost fifteen dollars to clean and rate it. Fifteen dollars! For sufficient reasons I left that timepiece at home.... (42–43)

In *The Last Voyage of Somebody the Sailor*, Somebody's watch becomes a "magic amulet" (284); while the watch is not strictly necessary for navigation, it is very close to being so: a relatively inexperienced navigator, for instance, with "that magical last-mentioned instrument, [can] ... make navigational calculations more accurate even than [the master navigator]

Mustafa's" (423); "thanks in part to my watch," says Somebody, "Shaitan was better sailed and ... better navigated" (496) than even Sindbad's best ship. The instrument and the attitude that Slocum calls "newfangled," then, are exactly that—"newfangled" is not strong enough a term—in Barth's novel. Somebody's watch acquires immense importance for navigation in this novel:

> Consider this, Brother. What you take to be an amulet may in fact be a key: the key to my returning to where I came from. I've yet to find the door, not to mention its lock, but with the key perhaps I can at least sail off in search of them. (380)

The image of Joshua Slocum, veteran of a solo voyage around the world, sailing into the path of a steamer in the night and disappearing without a trace is a powerful one. The metaphoric pull of the single-handed home-made sailboat being run over by indifferent western steamers—the precise vehicles that Somebody tries so hard to spot for comfort during his first days in Sindbad's world—is very strong;[15] the reference to Slocum is a resonant one.

E. *The Flying Carpet*

If the reference to Joshua Slocum in *The Last Voyage of Somebody the Sailor* is a resonant way to end the novel, so the reference to Richard Halliburton's adventures is an equally resonant way to begin it. "An airplane pilot is what I was going to be," notes Simon; "a *famous* pilot" (29, 50). Simon's ambition to be a "famous pilot" arises in part from his reading: "I was," he says, "a great fan of the adventurous Richard Halliburton, who had personally autographed Sam Moore's copies of his travel books" (87).

The web of reference to Halliburton becomes a very intricate one, tracing, as it does, the twin worlds of reality and romance, and the disjunctions between them. For although Simon's infatuation with Halliburton's real-life romances initially resembles Don Quixote's fascination for romance literature, the conclusions they come to are very different. Cervantes notes this about his favourite knight:

> he passed his time in reading books of knight-errantry.... In fine, he gave himself up so wholly to the reading of romances, that a-nights he

> would read on until it was day, and a-days he would read on until it was night; and thus, by sleeping little and reading much, the moisture of his brain was exhausted to that degree, that at last he lost the use of his reason. A world of disorderly notions, picked out of his books, crowded into his imagination; and now his head was full of nothing but enchantments, quarrels, battles, challenges, wounds, complaints, amours, torments, and abundance of stuff and impossibilities; insomuch, that all the fables and fantastical tales which he read seemed to him now as true as the most authentic histories. (Cervantes 4–5)

The general form of Cervantes's novel echoes throughout *The Last Voyage of Somebody the Sailor*. Simon's romances take over his imagination so completely that before long he is in the middle of one of them: "Our setting is a Baghdad evening; we're coming onto an Arabian night" (18).

References in *Somebody the Sailor* to Halliburton's adventures are very specific. They trace the same disjunctions, but where Don Quixote fades into his fantasy world, going mad in the process, Simon notices the differences and chooses to change his dreams. The planes in the two books, for instance, initially have striking similarities; they turn out to have profound differences. "Howard Garton," notes Simon, "led us through the small, empty, dirt-floored hangar behind his office and out to a Bellanca four-seater, red and yellow like his clothes or Aunt Rachel's Bon Ami can" (51). Halliburton's plane, on the other hand, is not described in terms as mundane as a "dirt-floored hangar" or a can of cleanser; it "hurdled the clouds, soared over the mountaintops, dived toward the sea, and skimmed the waves. Two sets of gleaming golden wings extended on either side of the scarlet body.... The wings were gold, the tapering body lacquered red. It was love at first sight" (1).

In Halliburton's account, the plane is romanticized. It is "gleaming gold" and "scarlet"; it hurdles, soars, dives, skims. In Simon's description, on the other hand, the plane is (simply, realistically) "red and yellow"; during the plane ride, furthermore, "things were exactly as I had known they would be" (53). The plane itself is "not all that different ... from one of Joe Junior's balsa-and-tissue flying models" (52). If "it was love at first sight" of the plane for Halliburton, it was disillusionment at first ride for Simon. Simon admits to "exhilaration" at the flight itself (55), but when his reality meets his fantasies, the fantasies, in effect, disappear. Noticing the difference between "reality" and romance makes him feel "taller and more serious forever" (55).

It is "an insufficient destiny" to fly planes, decides Simon; even though planes are magical in Richard Halliburton's adventures, even though in those adventures they are named "The Magic Carpet" (1), the reality of planes themselves is rather mundane. They are lacking in imagination. Simon realizes this as he realizes that "we didn't need helmets and goggles" (53); that is, he realizes that the accoutrements of romance are unsuited to reality, and reality is not enough for Simon. Later in life, Simon goes "to Baghdad ... not on a flying carpet but inside a tightly rolled one, by upstream barge and donkey-back ..." (517).

These differences can also be read allegorically, just as Simon's decision, made over the course of his novel, that reality and realism are "not enough" unless spiced up by Halliburton's brand of romance, can be read allegorically, as well. Halliburton's plane is named the Flying Carpet (1); Simon concludes with a very quick and cursory experience with Halliburton-like adventures that this was "an insufficient destiny"; he concludes, allegorically, that he cannot write only travel books, though that is what Baylor does, initially.

The twin images of Slocum and Halliburton must have been irresistible to Barth. Halliburton, sailing from the Orient to San Francisco, also stumbled into the path of a steamer and disappeared at sea: he "perished while sailing the Chinese junk Sea Dragon across the Pacific towards San Francisco"; his last communication was transmitted 24 March 1939 (579).

F. *The Sindbad Voyage*

The reference to Tim Severin's *Sindbad Voyage* is an explicit reference of a slightly different kind from some of the intertexts that I have already cited. Severin is an adventurer who has made several "authentic" voyages—replications, really, of historical adventures. In 1980, Severin decided to replicate the famous voyages of Sindbad. It is interesting how closely bits of Barth's novel correspond to Severin's account. One of the more memorable references is the bartering and discussion of bartering by Sindbad himself for "first-quality poon logs from Beypore" (73). Poon is considered to be the perfect wood for ship spars; compare Severin's account of negotiating for poon with Indian wood dealers (42–43). One of the more humorous references, on the other hand, is Barth's comment about imagination and fiction and Severin's use of them: "experience has taught him [Severin] that the essential moments of such replicated adventures are those when nothing in sight suggests the Here and Now" (325). Severin's is an actual trip; Barth's is a fictional account. But by including Severin's adventure as part of his novel, Barth manages to include yet one more action-adventure yarn from seafaring lore in his own framework. He manages thereby to stretch the boundaries of where novels in general, and his novel in particular, can be situated.

G. *Other references*

It is an odd Barth story, indeed, that does not have an appearance in it of one or several of Borges's stories. Tobin has pointed out the importance of both "Zahir" and "The South" to *The Last Voyage of Somebody the Sailor*; also implicit in the welter of allusions, quotations, and references in this novel is Borges's story "Pierre Menard, Author of the Quixote," whose "implicit theme ... is the difficulty, perhaps the unnecessity, of writing original works of literature" ("The Literature of Exhaustion," *The Friday Book* 69).

When Simon disappears from one world, he reappears in another. For instance, when he leaves Sindbad's world towards the end of the novel, he wakes "once again naked, salt-soaked, sunburnt; again sore-muscled, ache-jointed, daze-headed, also ungroomed unbarbered unoriented and all but out of breath" (*The Last Voyage of Somebody the Sailor* 570). This description is strongly reminiscent of Homer's description of Odysseus at the end of Book V and the beginning of Book VI of *The Odyssey*: there, Odysseus was "naked.... [And] terrible in their eyes, being marred with the salt sea foam" (90). More important, however, is the reference to Barth's own *Tidewater Tales*, specifically to a passage in which several of the characters tell their favorite stories; one character's favorite is the story of "poor ship-wrecked Odysseus—naked, sea-grimed, exhausted from swimming, and alone" (179). Barth's reference thereby situates this novel not only in the canon of world literature, but in the canon of Barthian literature, and in the canon of literature that cites classical literature, as well.

"Somebody's first voyage," if one can call it that, begins in the way that western stories often do: he uses the time-honoured practice of beginning "Once upon a time" (27). This is an immediate signal that the story that is about to appear, despite its title, does not arise from the orient at all, is not eastern in origin, but is very much a traditional western story. The story continues, not exactly the way one would expect a fairytale to continue, but slipping into the rhetoric of biography and autobiography, because not only is the story predominantly in first person, but it also includes lines from standard biography and autobiography: "Like many another solitary though not unhappy child, young Simon William Behler invented playmates" (27). This sort of story is standard practice in the West. Somebody's audience, however, is offended. Patricia Tobin notes the reason: "the 'Zahirites' were a medieval religious sect" who abhorred invention and imagination. Thus "the Muslims' negative critique of Somebody's tales had a religious basis in the generalized taboo on Arabic autobiography: That one must not add anything to the Koran is a Zahirite prohibition" (172).

The beginning of Barth's novel slips in time, as well, becoming, like *Tristram Shandy*, a mock epic autobiographical beginning. Somebody's story begins in the womb, while Shandy's begins with his conception (this is also a reference to Barth's own "Night-Sea Journey," no doubt, with its voyaging sperm).

H. *Reading, Writing, and Sex*

The first element to mention in this section is the tension between the need for originality and the impossibility of ever being original. The problem becomes how to recycle an old story so that it seems new: this begins to take the form of exhaustion and renewal, which have been central concerns of Barth for some time. Many connections are possible from here to classical stories, including mythology, and certainly including stories like the 1001 Nights. This tension is at the root of the most frequently recurring metaphor in the novel: the image of the virgin. There are virgin storytellers, virgin brides, virgin listeners; it is obvious that "maidenhead" is an illusion, a constructed state of mind, really, an agreement between negotiating parties, and only ever a convention. The maidenhead as it appears in this novel is frequently meticulously reconstructed; in any case, maidenheads are purely negotiated, agreed-upon contracts that have little, if any, basis in reality, though they are, apparently, prized above anything else in the entire novel: they are, in short, a convention and a contract, nothing more. They are a public acknowledgement of artifice and dissimulation. This first concern is related to a second, which is the relationship between tradition and innovation. The novel opens with the promise that the audience has never heard this story before; yet the entire novel is made up of stories that we have, in fact, heard before. Even the promise that we have not heard the story of the real death of Scheherazade, which will be told in the work that we are now reading, is itself something that we have heard before. All of this entails the asking of the question of how to recycle an old story so that it becomes, or at least seems to become, new.

The sexual aspects of this novel belong to a line of argument that Barth has used before. In a sense, this, too, is intertextual; for instance, the climax of stories and the climax of sexual orgasm was

> a favorite ... [that] would lead them to a dozen others between narrative and sexual art.... writing and reading, or telling and listening, were literally ways of making love. (*Chimera* 24)

One should note also the

> similarity between conventional dramatic structure—its exposition, rising action, climax, and denouement—and the rhythm of sexual intercourse from foreplay through coitus to orgasm and release. Therefore, also, they believed, the popularity of love ... as a theme for narrative. (*Chimera* 24–25)

Further,

> [1001 nights is] the best illustration of all that the relation between teller and told was by nature erotic. The teller's role, he felt, regardless of his actual gender, was essentially masculine, the listener's or reader's feminine, and the tale was the medium of their intercourse.
>
> "That makes me unnatural," Sherry objected. "Are you one of those vulgar men who think that women writers are homosexuals?"
>
> "Not at all," the Genie assured her.... he had not meant to suggest that the "femininity" of readership was a docile or inferior condition.... Narrative, in short ... was a love-relation, not a rape: its success depended upon the reader's consent and cooperation, which she could withhold or at any moment withdraw; also upon her own combination of experience and talent for the enterprise, and the author's ability to arouse, sustain, and satisfy her interest—an ability on which his figurative life hung as surely as Scheherazade's literal. (*Chimera* 25–26)[16]

In addition, tales of loves lost or loves found or loves, at least, misconstrued, are important to note here. There are at least two works of F. Scott Fitzgerald that appear in *The Last Voyage of Somebody the Sailor*. First, Daisy has a namesake in *The Great Gatsby*, but she also, we are told, "affirmed her solidarity with the resident ghost of Zelda Fitzgerald while lamenting that 'Persimmon' had not been her F. Scott" (305). In addition, there are smaller references: Sahim asks Simon, for instance, "whether [he] could see a tiny light low on the northeast shore, like a fallen star of faint magnitude" (511), which echoes the light at the end of Daisy's dock in *Gatsby*. There are other similarities to *The Great Gatsby*: with the East/West Dorset split and the yearning after the unattainable Daisy who lives out of reach in West Dorset echoing East and West Egg and a yearning Gatsby from Fitzgerald's novel; there is even and echo in Simon's needing to become a Somebody before he deserves Daisy (86).

And *Gatsby* is not the only Fitzgerald novel to make an appearance in Barth's book. Julia notes of Daisy, for instance: "after Dad's death she gave me chapter and verse: the whole *Tender is the Night*" (316).

Fitzgerald writes:

> After her mother died when she was little she used to come into my bed every morning and sometimes she'd sleep in my bed. I was sorry for the little thing. Oh, after that, whenever we went places in an automobile or a train we used to hold hands. She used to sing to me. We used to say, "Now let's not pay any attention to any body else this afternoon—let's just have each other—for this morning you're mine." A broken sarcasm came into his voice. "People used to say what a wonderful father and daughter we were—they used to wipe their eyes. We were just like lovers—and then all at once we were lovers." (*Tender is the Night* 18).

Nicole in *Tender is the Night* refers to herself as "Daddy's Girl" (227); Julia Moore smiles that her heart "belongs to Daddy" (319). Even Bon Ami, the cleaning powder that cycles through Barth's novel, is mentioned by name in *Tender is the Night* (178).

In a novel so redolent with sexual relationships, on one hand, and so many written intertextual references on the other, it is no surprise that incest appears as a theme: intensive quotation has long been linked metaphorically to incest. It is interesting to see how that theme is handled in this novel; it is important to note that it is not the sex itself, but the exploitation connected to incestuous sex, that is problematical.

Julia says:

> I think I think that the sexual exploitation of human beings is a crime against humanity and that as charming as he was about it, my late father's incest streak was unpardonably exploitative, to put it mildly.... Incest is a loaded word.... Exploitation is the crime. Sam should've kept his hands off us until we were our own women. Experienced. Independent. How we handled him then would've been nobody else's beeswax. (317)

If this can be read allegorically, it can be seen as a defence of the kind of very intricate, very sophisticated, and very extensive quoting of the work of others that is practised in this novel. The quoting is, as I have already mentioned, a sort of incest; but the incest is not exploitative if all the parties involved are mature, experienced, and independent.

The sexual metaphor and the intertextual references and the various other textual ploys that I have mentioned here all add up to the structure of this novel, which is fundamentally gamelike: the boundaries are set, not in advance, and not through explicit articulation; but just as surely, and just as firmly, the structure of the novel forms into what is recognizably a playing field with very definite limits and bounds.

3. Conclusion

The Last Voyage of Somebody the Sailor is playful, as are most of Barth's writings. It plays with language; it plays with references; it plays with sex; it plays with story; it plays with genre. It is, at a profound level, a mix of other texts and a mix of other genres, including many seafaring adventure yarns, and not just a novel that includes these references. Part of Barth's novel is relatively straightforward and traditional Western realistic narrative, while the rest of it is an Eastern fantasy tale that is populated by genies, rocs, giant serpents, and monstrous ogres which belong only in the most fantastic of fables—in the pages of *The Thousand Nights and One Night*. And while Barth's novel is a combination of those opposites, it is also an amalgamation of short stories, other novels, poems and songs about story telling, and real accounts of seafaring adventures.

The Last Voyage of Somebody the Sailor is thus structured as a web of reference and counter-reference, of very obscure and equally obvious citations. This merging of opposites, this blending of genres and genics, this precarious negotiation of boundaries, is what makes Barth's novel a novel; it is what, finally, gives it its structure and shape; it is what makes it literature; and, not incidentally, it is what makes it game-like. It is the constant, playful, gaming references to other books, other novels, other literature, that makes the establishment, negotiation, and control of boundaries not only a structural feature of *Somebody the Sailor*, but its primary thematic concern as well. Other books; intertextuality; reference points in the sea of story: the limits which those references represent are the limits of game; they are the walls of a literary labyrinth.

The intertextuality of Barth's *The Last Voyage of Somebody the Sailor*, finally, cuts two ways: on one hand, the intertexts specify, narrow, and define this novel. They offer the familiar comforts of western autobiographical realism, and they offer the fantasy of eastern story cycles. The intertexts provide contexts and give the novel its distinctive shape, texture, heft. On the other hand, they expand the limits of this novel to take its place in the Sea of Story, in the wider tradition of story-telling, of seafaring adventure, of history, myth, legend. Finally, they provide the necessary structure inside of which Barth and Barth's readers are free to play.

CHAPTER FIVE

Dancing Between Worlds: Louise Erdrich's Complex Playground

Perhaps if I invoke Clare, the patron saint of television.
(Erdrich, *Baptism of Desire* 3)

Tracks is a profoundly divided text, articulating as it does on the one hand the already deeply divisive, historically "real" Chippewa[1] experience, while at the same time displaying a playfully postmodernist textual self-reflexivity.[2] It thereby reflects and articulates the divisions and the borders that I have been discussing as typical of postmodernism, while taking part in the play that is also typical. As Erdrich herself has pointed out, *Tracks* has "two narrators who tell the same stories in different ways" (White and Burnside 167). What makes this novel gameful is the structure that allows for, or in fact demands, alternative possibilities, inside a framework that allows play while still limiting it.

The divisions in *Tracks* are many and varied: they are the divisions, for instance, not only between American Indian and white, a split reflected in Erdrich's own genealogy,[3] but also, and often more profoundly, between the oral and the written, between story and history, fiction and life, dreams and reality, myth and religion. These divisions are expressed through the novel's two narrators, two peoples, two cultures, and in its multiple and multiplying story-lines. The novel expresses these divisions without expressly choosing one or the other. There are, in fact, at least two novels at work here: one is a realistic, fairly straightforward historical novel which tells a Chippewa story from a Chippewa point of view, more or less; the other is mythic, told by a traditional Chippewa teller. The trickster, Nanapush, traditional Chippewa storyteller, sits on one hand; Pauline, a so-called "half-breed" Chippewa by birth, but during the novel increasingly a mad religious zealot, sits on the other. *Tracks* does not, strictly speaking, straddle the line between pairs of options; it does not walk the border between archives, but delves into each. *Tracks* hesitates, invites, proffers, and withholds information to and from its readers, demanding that they participate in the experience of fiction. *Tracks* assumes, in short, that readers will play the game of fiction, resolving or holding in suspension those elements of the story that are not resolved, and making the choices that will resolve other elements. It offers differing, competing, in some ways contrasting views of the same issues and events, and challenges the reader to decipher them.

1. On One Hand, Realism

The first novel, the realistic one, has its beginnings in historical records and interpretations of Chippewa experience: the elements of the "real" that appear in *Tracks* have their beginnings in history. E.J. Danziger, in his *The Chippewas of Lake Superior*, is interested in the impact of European and American societies on the culture of the Chippewa. Danziger's assessment of the Chippewa at the beginning of the twentieth century, the time period of *Tracks*, is that they were "a people of two worlds," caught in the midst of profound changes that were happening largely without their consent and largely beyond their control (109). They were trying, on the one hand, to maintain as much as possible of their traditional ways of life; and on the other, to cope with, and adapt to, an encroaching white culture voracious for land. Danziger compares data from U.S. federal documents and statistics compiled by the Bureau of Indian Affairs (BIA),[4] with changing traditional Chippewa values and beliefs; he concludes that the BIA in effect set out deliberately to destroy the Chippewa way of life by enforcing white cultural values at the expense of Indian traditions.

Danziger notes, for instance, that "until the 1930's the federal government's primary goal for the reservation Chippewas was 'civilization' (or acculturation), meaning eradication of the most apparent features of the Chippewas [sic] traditional culture and their replacement by such white cultural traits as the work ethic in reference to cultivating the soil. Once this was accomplished, 'assimilation' (or social integration) and economic self-sufficiency would logically follow" (92). In other words, once the Indians had first become dependent on an economy that relied on their role in the fur trade, and then on government annuity money from their land holdings,[5] they were to be weaned from receiving government money: they were to be forced into becoming "civilized" farmers. But in order for them to be farmers, their fundamental ways of living had to be changed. To measure the effectiveness of various efforts at "civilization," Danziger notes, the BIA systematically collected such economic statistics as data on "land ownership, family incomes, occupation of the head of the house, and seasonal sources of sustenance.... [As well as] educational facts on Indian enrolment, dropout rate, and achievement tests [sic] scores" (93).

The data gathered by the BIA are very informative for their remarkable cultural biases and ignorance. First, the Chippewa had no tradition of land ownership, and would have had very little idea of what it would mean to "own" land.[6] The answers to questions posed to a white settler or farmer about land ownership would have helped to measure the success of the settler or farmer; the Chippewa's answer essentially would record who had managed to sell or not sell an allotment. It is a question of acquisition on one side, and relinquishment on the other, and thus the data become only a (partial) measure of the success of the BIA's programs, while yielding little or no information about the Chippewa themselves. It is informative that Nanapush in *Tracks* complains that "the land was snatched from under us at every step," that it became a "land we would never walk or hunt, from which our children would be barred" (4, 174). Nanapush does not complain that the land has been stolen from him, or that it has been snatched out of his hands, or that he suddenly does not own it, or that he is no longer free to farm it: the land has not been stolen from him, but has been snatched *from under him*, making it impossible for him or his children to walk or hunt on it. Nanapush complains because he suddenly finds himself and his people without a way of life: his are the Indian terms, and they are terms that have nothing whatever to do with farming or ownership.

Second, for the traditional Chippewa, "family income" was problematic in at least two ways, not only regarding the definition of "family,"[7] but also touching the notion of "income." The Chippewa concept of family, first, involved an elaborate clan system. The traditional Chippewa family was a wide-reaching network of relationships that corresponded exactly neither to the white notion of the "nuclear" family, nor to the white "extended" family. Clan members did not necessarily have a blood relationship to each other, though they might have; and the lack of a blood relationship did not lessen the importance of a particular family or clan connection.[8] Some of these complexities are addressed in *Tracks*: "Sophie and Clarence Morrissey both married Lazarres. Some said those were forbidden partners, cousins in their own clan, but others maintained we were so mixed in with French now it didn't matter. They went to be married at a distant church in Canada, where the fact that they were related was not mentioned or known" (179). It appears that the Morrisseys and Lazarres are clan cousins; it is not clear whether they are blood cousins by white reckoning. This apparent lack of a blood relationship does not, however, make the marriage any less incestuous in the Chippewa world, even though it is legal in the white one. The marriages represent only one of the ways in which white influence in *Tracks* erodes Indian culture from within.

In addition, asking about "income" in what was still, at the time, a semi-nomadic hunter-gatherer culture is both ludicrous and confusing. There is an interesting exchange in *Tracks* marking the moment at which the Indian people know that Fleur has decided to make her cabin beside Matchimanito a more or less permanent home: "That's how everyone knew she had come back to stay. It was the money. She paid the annual fee on every Pillager allotment she had inherited, then laid in a store of supplies that would last through winter. And it was the money itself, the coins and bills, that made more talk. Before this, the Pillagers had always traded with fur, meat, hides, or berries" (36). In other words, the very idea of a Pillager having an "income" rather than dealing in barter goods was a source of fascinating gossip, not the answer to a mundane question originating in a poll.

Third, the question about the "occupation of the head of the house" is equally problematic. While the "family" (however it is defined) was a very important social unit for the Chippewa, the white "nuclear family" with its (usually male) "head of the house" was not the Chippewa norm. This question prompts more questions than it answers: was the so-called "head of the house" to be defined as the individual who "brought home the family 'income'"? Does the question of "occupation" address the prowess of a hunter, for instance, or does it rather involve unspoken assumptions about "work" and "income," and other white biases? The question in general was unlikely to refer to the nets of influence extending through the matriarchal or patriarchal clan systems, or to address any specific concerns of the Indians themselves. In *Tracks*, who is properly the "head of the house" in the cabin by Matchimanito, Eli Kashpaw or Fleur Pillager? What could be entered in a poll about the "occupation" of either of them?

Finally, the questions about "sources of sustenance" and the patronizing tone of the declaration that schemes like allotment would make the Chippewa "economically self-sufficient" speak eloquently of the profound changes that had been instituted by the BIA. This issue, too, is addressed in *Tracks*, in Nanapush's eloquent declaration: "In late winter, the fish in the lake swarmed to the surface and we netted them through holes in the ice. Once we started eating food we caught for ourselves, my bad dreams stopped" (182). Though the Chippewa might not have been "*economically* self-sufficient" before the coming of the BIA, it is clearly the attitude of *Tracks* that they were self-sufficient in real terms, and would have preferred to remain that way.

In addition to economic statistics, the BIA was interested in (re-) education, and so kept track of the numbers of Indians involved in the school system: enrolments, dropouts, test scores were recorded carefully and systematically. From the point of view of the BIA, the government school system was positive, helpful, "civilizing," enlightening; from the point of view of *Tracks*, the opposite is true: "the young, like Lulu and Nector, return from the government schools blinded and deafened" (205). Considered unimportant for BIA fact-gathering were such factors as the bothersome "persistence of Chippewa social customs (clan relations, child rearing, courtship and marriage, village life, death and burial rites), religious thought and practice, artistic expression, and tribal political organization" (Danziger 93). *Tracks*, on the other hand, is precisely concerned with the provisional survival of those customs.

The major reason that such factors were deemed unimportant statistically by the BIA was that organisation's determination to change traditional customs by changing economics and education; it seems to have been assumed that "civilization" naturally would change the more "primitive" aspects of Chippewa life. Chippewa religious expression, for instance, was considered crude, at least by some:

> ... it may be said that the Chippewa recognized a Supreme Being to whom they offered reverence and worship. However, these warped ideas and superstitious practices made their cult almost unworthy of the name religion. Only the vivifying power of the Gospel of Christ could lift them from their degraded state. (Levi 37)

Pieces of Chippewa decorative arts were treated as articles of commerce by whites, and so became increasingly affected by white standards and taste, leading to a classic dilemma: should one privilege commerce or art? "The spirit of real art decreased as the imitation of nature increased until the floral patterns in use among the Chippewa of to-day have no artistic value"

(Densmore 183, 186). Finally, political assimilation into white systems of government was assumed by the BIA; Chippewa civilization was considered "primitive," and European-based governments "all agreed in assuming the right of dominion, based on discovery, without regard to the natives" (Hodge 1: 500; cf. 1: 498–503).

The fact that BIA statistics ignored social, religious, artistic, and political customs does not mean that the Chippewa were free of direct persecution on cultural and religious grounds. Sister M. Carolissa Levi is unintentionally ironic when she notes:

> Scarcely a historic site in the Northwest can be named which a voyageur or a trader did not reach before the "Black Robe" or the "Brown Robe," yet the majority of the former left but a slight impression upon their environment in comparison with the permanent heritage of spiritual values and practical culture bequeathed by the latter. (38)

It is informative that the most eager of recipients of "spiritual values and practical culture" in *Tracks* is Pauline, the part-time narrator of the novel and, as it turns out, the increasingly mad Sister Leopolda of Erdrich's other fiction. The Chippewa at the time of *Tracks* were indeed "a people of two worlds," striving to keep a dying culture alive, and striving at the same time to cope with a surging white culture determined to either change them or submerge them.

Tracks expresses this split and doubled world in several ways. First, the Indian/white clash over land and the Indians' reluctant though inexorable relinquishment of that land is the central plot element of the novel. The white consumption of Chippewa land happened at an alarming rate in the "real world" during the time period of, and just before, *Tracks*. Morison notes, for instance, regarding the Dawes Act of 1887:

> The act, in general, provided that the President of the United States should direct that a reservation be broken up when and if he had evidence that the Indians wanted it; then a homestead would be granted to each family, and the unallotted remainder of the reservation would be purchased by the government for sale to white men, the money to be put in trust for the tribe. After allotment began, in 1891, the acreage of Indian reservations was reduced 12 per cent in a single year. Congress then speeded up the process by passing another law which allowed the allottees to lease their lands. That really doomed the system. Indians

> living on a reservation lapped about by white men's farms, faced with the alternative of becoming a tribal slum on the prairie or unwilling homesteaders, snapped hungrily at the allotment bait, knowing that individual farms could now be leased, and hoping to live well on the rent. In 1894 ... the Omaha and Winnebago in Nebraska had leased lands to a real estate syndicate for 8 to 10 cents an acre, which the syndicate released to white farmers for $1 to $2 an acre, per annum. Out of 140,000 acres allotted by 1898 to those two nations, 112,000 had been leased.... In the half-century after 1887, Indian holdings decreased from 138 to 48 million acres. (753–54)[9]

And even more pertinently, Weil writes: "Within three weeks of the passage of the 1906 legislation [the Clapp Amendment], 250 mortgages against reservation lands [at White Earth] were recorded. Those who took them out had absolutely no training or help from the government in how to retain or use their money, and most was soon wasted. By 1910 three-quarters of the allotments had been sold, and logging was rapidly destroying White Earth's pine forest" (78–79). It is thus not surprising that *Tracks*, set in the years between 1912 and 1924, is dominated by questions of trees vs. land, farming vs. hunting, and "tradition" vs. "progress." *Tracks* is thus at least in part a piece of historical realism, a straightforward novel "about" land claims and jealousy and corruption, the clash between Indians and whites, and the problems that result when the two cultures meet.

However, Erdrich has not left herself open to charges of accepting easy solutions or indulging in Hollywood-like sentimentality by making the distinctions as stark as that sounds.[10] It is true that some Indians are exploited, innocent, and foolish victims at the beginning of the novel: "Starvation makes fools of us all," notes Nanapush. "In the past, some had sold their allotment land for one hundred poundweight of flour" (8). And it is true that some stay foolish throughout:

> "Look, fool," I [Nanapush] said. "Open your eyes. Even your baby brother has a better grasp of what is going on. We're offered money in the agreements, cash for land. What will you do with the money?"
>
> "Right now?" Eli asked in a belligerent, stalling way.
>
> "Yes," I said, "what would you do with fifty dollars this moment?"
>
> "I'd drink it up," he said in a pouting voice, daring my wrath even though I knew he rarely drank. I gave him no satisfaction, just kept the argument going.

> "Like many," I said, "you'd wake with no place to put your foot down." (98–99)

But if it is true that there are confused and exploited Chippewa, it is also true that there are Indians in this novel who are very conversant with "the system," and able and willing to use it to their own ends. The Morrisseys, for instance, are "our best farmers" (37). They are very skilled in government subtleties: "They were well-off people, mixed-bloods who profited from acquiring allotments that many old Chippewa did not know how to keep. [Their] farm was big for those days, six hundred and forty acres" (63). And they seem able and willing to adapt to new ways of life; they "had signed the new purchase agreement with the Turcot lumber company, and now spoke in its favor to anyone whom they could collar. They even came to people's houses to beg and argue that this was our one chance, our good chance, that the officials would drop the offer. But wherever Margaret was, she slapped down their words like mosquitoes" (111).[11]

Tracks, then, is no romantic tale of the "noble savage" being exploited, and tragically but heroically being conquered. What begins as the struggle by some to retain the "old" ways against white domination resolves itself into a struggle among Indians: between those who cling to tradition and those who accept, adapt to, and even manage to exploit the inevitable changes that were originally the fault of the whites. This struggle is enacted in several places in the novel, including in the rivalry between Nanapush and the Morrisseys, or Margaret and the Morrisseys, or even, more broadly, in the general scheme of Nanapush's story to Lulu; but it is perhaps represented best by Fleur Pillager, "the funnel of our history .... the lone survivor of the Pillagers" (178), a woman who believes "no one would be reckless enough to try collecting for land where Pillagers were buried" (174), who is confident that no one would "dare throw us off the shores of this lake" (175), against Nector Kashpaw, the young and visionary Indian who "paid the money down on Kashpaw land from foresight, shrewdness, greed, all that would make him a good politician" (209). Nector "wins" because he manages to keep his land by using Fleur's money because she is foolish enough to trust him to make her payments for her. In this struggle, it becomes sadly obvious that Margaret is right when she observes that Fleur is "living in the old days when people had respect" (174). Fleur *is* living in the past: she refuses to adapt; she refuses to recognize that a profound split has already appeared in her community.

Tracks is a record of that split. It is the story of a division; it becomes a novel caught hesitating between the physical and spiritual and social world of the Chippewa and the abstract world of the whites, between an actual way of life and a general philosophy of being. From this point of view, it becomes informative and worthy of note that the white forces that drive the changes taking place in the novel have no real identity. The whites have no faces; and even more important is what Weil calls "the deep social division within the tribe between the 'mixed'- and 'full'-bloods" (71).

The Turcot Lumber Company, for instance, is emblematic of corporate white America, and thus of white "progress." It is a dominant, yet shadowy, force which has "very kindly ... consented to start the lumbering operation on the far side of the lake.... [giving] the residents time to gather their possessions. Even time to build somewhere else" (208). And the logging company is not alone; there are strong rumours of other requests for land, without actual requests or physical requesters: "Father Damien had already heard.... 'There's some who want to build a fishing lodge,' he said in a gentle voice. 'They're willing to trade for an allotment someplace else'" (175).[12]

The whites are represented in *Tracks* by the dollar bill as the unit of exchange, replacing Indian forms of barter. The whites are indecipherable images on paper ("annual fee lists, foreclosure notices" [173]); colours on a map ("a pale and rotten .... lapping pink" [173–74]); shadowy forms and sounds across the lake: "It began as a far-off murmur, a disturbance in the wind.... Ringing over the water and to our shore came the shouts of men, faint thump of steel axes. Their saws were rasping whispers, the turn of wooden wheels on ungreased axles was shrill as a far-off flock of gulls" (206). The whites are an elusive, yet ubiquitous and inexorable enemy that manages to portray itself as no enemy at all: "And yet," notes Nanapush, "I learned the Agent was not against us. I entered his office and stood before his desk. I was told that it was not his fault the trees were sold and cut down. Nor was the tribe to blame. There was no adversary, no betrayer, no one to fight" (207).

The whites finally exist purely as an abstraction. They are legislations and corporate identities, and they win the battle with the Chippewa through managing subtle yet profound changes to Chippewa habits, expectations, and ways of thinking: "Our trouble came from living, from liquor and the dollar bill. We stumbled toward the government bait, never looking down, never noticing how the land was snatched from under us at every step" (4); "dollar bills cause the memory to vanish, and even fear can be cushioned by the application of government cash" (174). In short, white culture is the destructive abstract reality that clashes with and finally overcomes concrete Chippewa reality, not least in the form of the "mixed-bloods" who exploit the rules so successfully to their advantage.

As the novel comes to its end and Nanapush finds his battle lost, he is forced to admit that he has been using faulty tactics. He believes he had had the right idea to begin with: "As a young man, I had made my reputation as a government interpreter, that is, until the Beauchamp Treaty signing, in which I said to Rift-in-a-Cloud, 'Don't put your thumb in the ink.' One of the officials understood and I lost my job" (100). That kind of direct subversion is the kind, he believes, he should have pursued. He should have engaged the fight on his enemies' ground: "I now saw what Father Damien read, looking into a distance I could not have imagined. He was right in that I should have tried to grasp this new way of wielding influence, this method of leading others with a pen and piece of paper. I looked around. If I had, perhaps the road I walked would not have been rutted by the wheels of laden wagons" (209). While *Tracks* can be seen as itself a paper subversion, it is informative nonetheless that Nanapush "sees," while Father Damien "reads"; one wonders whether Nanapush has, in fact, learned anything.

Finally, Nanapush's narrative is ostensibly an oral, not a written one; it is a story told to Lulu, one meant to preserve her history and to advise her of appropriate decisions for the future. *Tracks* in essence contains the record of the story: the novel is, literally, the story's tracks.

If this were all there were to *Tracks*, it could profitably be read as essentially a more or less realistic, more or less accurate historical novel, a moderately fictionalized account of Indian history in general and of the Chippewa experience in particular. In fact, this is the tack taken by most of the critical writings on *Tracks*, and on Erdrich's fiction in general. See, for instance, Peterson's "History, Postmodernism, and Louise Erdrich's *Tracks*"; Stripes's "The Problem(s) of (Anishinaabe) History in the Fiction of Louise Erdrich"; Maristuen-Rodakowski's "The Turtle Mountain Reservation in North Dakota: Its History as Depicted in Louise Erdrich's *Love Medicine* and *Beet Queen*." However, that is neither the only kind of writing, nor the only novel that makes up *Tracks*.

2. And on the Other Hand, Relentless Self-Referentiality

Sometimes Erdrich's style and preoccupations with concerns other than "traditional realism" have been received warmly: see Catherine Rainwater's "Reading Between Worlds." On other occasions, her style has resulted in harsh criticism, such as that in Leslie Marmon Silko's "Here's an Odd Artifact for the Fairy-Tale Shelf": "Erdrich's prose is an outgrowth of academic, post-modern, so-called experimental influences.... Self-referential writing has an ethereal clarity and shimmering beauty because no history or politics intrudes to muddy the well of pure necessity contained within language itself" (179).

The second novel, the one that concentrates on the postmodernist split and doubling aspect of *Tracks*, is expressed in its self-reflexive metafictional trappings. The dedication, first of all, is to "Michael,"[13] and reads:

> The story comes up different
> every time and has no ending
> but always begins with you.

Even though the dedication is not, strictly speaking, a part of the novel, it is nonetheless a part of the novel that announces itself from the beginning as a story that is shifty, changing, changeable, and circular.

Furthermore, not only the story "to Michael" is circular, but the novel itself is also circular. It is a circle that involves Lulu and Nanapush, and her story, and his. From the first page, this is a story directed at "Granddaughter" (1), someone whom Nanapush addresses as "My girl" (2); yet that means that it is a story told to the young woman who emerges from "the rattling green vehicle the government sent" (226); Lulu steps from the bus, some years *before* the talk from Nanapush that is this novel begins, yet several years *after* the story that it tells begins:

> But your grin was bold as your mother's, white with anger that vanished when you saw us waiting. You went up on your toes, and tried to walk, prim as you'd been taught. Halfway across, you could not contain yourself and sprang forward. Lulu. We gave against your rush like creaking oaks, held on, braced ourselves together in the fierce dry wind. (226)

Circular narratives are a staple of postmodernism, but they are also part of the tradition of Indian storytelling that is also, in turn, such a vital part of what this novel is. Paula Gunn Allen writes:

> The way of the imagination is the way of continuity, circularity, completeness. The way of the intellect is the way of segmentation, discontinuity, linearity. We persist over time; we endure. We forget our origins and lay waste to the claims of the past, simultaneously deeming them to be the only truth, but not overtly. ("Bringing Home the Fact" 563)

The differences in traditions are reflected again immediately in the various methods of chronology that are represented on the first page. First, we are told that this is Chapter 1: this notation is a secure part of the tradition of the English novel. Next, the English season and year are "Winter 1912." Again, this is strictly traditional to the English novel. Then comes the Chippewa season: "Manitou-geezisohns," which is, translated, "Little Spirit Sun"; our narrator is "Nanapush." The novelistic conventions of providing time and place—setting—are followed, but they are immediately undercut, and in need of re-reading when the reader finds that although the English season is given, and a translation is provided from the Chippewa for the alternative name of the season, the translation appears to be a courtesy to the reader. This occurs because this first narrator, after all, is Chippewa. One effect of all this notation is that "Winter 1912" is recast to take into account a Chippewa narrator; the astute reader may infer a good deal from that date to do with land claims, and so on; even the less astute may infer a good deal about context from these few short words.

The divisions that I have begun to articulate are in the novel from beginning to end, but are suggested perhaps most profoundly by the way the novel is narrated. Neither of the two narrators is objective; neither even pretends to be so. Both utilize first person narratives, and so both have a vested interest in the story; neither is, therefore, entirely trustworthy. Unreliable narrators are not new, but in this case, Erdrich does not give us a solid base that would make it possible to make decisions about what "really happens" in the novel, or about which narrator we should choose most to believe; the narrative of each narrator is already explicitly in each case an interpretation, one that may be accepted or rejected, one that may need revisions, one that is necessarily in its turn going to be subjected to further interpretations imposed by readers.

A. *Nanapush*

The first narrator is Nanapush.[14] He is male, a skilled story-teller, the namesake of a mythical Chippewa figure, and his story, despite its initial novelistic trappings, is apparently an oral one, told to Lulu, with the intent of imbuing her with tribal knowledge. It is designed to convince her not to marry into the Morrissey clan of mixed-bloods: "Lulu, it is time, now, before you marry your no-good Morrissey and toss your life away, for you to listen" (218).[15]

Nanapush is not only Lulu's grandfather, but we know from his name that he is a traditional tribal wise man. He gives this account of his naming:

> My father said, "Nanapush. That's what you'll be called. Because it's got to do with trickery and living in the bush. Because it's got to do with something a girl can't resist. The first Nanapush stole fire. You will steal hearts." (33)

Nanapush in Chippewa legend is the trickster, the storyteller, the seducer, the lover of women. Alan Velie writes:

> The tribal trickster is not a single figure; tricksters differ greatly from tribe to tribe and even from tale to tale in the repertoire of the same tribe.... Whatever his form, trickster has a familiar set of characteristics: he plays tricks and is the victim of tricks; he is amoral and has strong appetites, particularly for food and sex; he is footloose, irresponsible and callous, but somehow almost always sympathetic if not lovable. (122)

This is, of course, as many critics have pointed out, quite a precise description of the Nanapush of *Tracks* as well as of the traditional figure.

Perhaps the most notable of the tricks that Nanapush plays in this novel is played on Pauline, who has, to her horror and dismay, admitted that it is one of her pledges to abstain from performing regular bodily functions as needed, but only under her strict discipline. Pauline notes, "my most secret practice ... was to allow myself only two times of day for that function [urination], dawn and dusk" (147). The trick that he plays is bawdy, inventive, somewhat malicious, but clever; it is utterly typical of his character and a representative example of his sense of humour. He decides to embarrass her by making her urinate. He begins by tricking her into drinking too much sassafras tea; the trick ends this way:

> Between his fingers, he was holding what the men down in Argus called a safety. He began to fill it with sassafras tea, from the spout. Before my eyes, the thin skin elongated and ballooned. I was sick. I began to shake all over, groaned deep in my chest.... Nanapush poured more tea. The others howled and rocked.... Then the skin burst and a wave poured across the table under Margaret's amused scolding. I said a thousand prayers in one drenched second. I made for the door in a crouching run and didn't care if I failed in the test, or even if I had to suffer a million years of the devil's laughter, just so I could relieve my burden. (151)

Nanapush represents, and tells stories that represent, the "old ways": myth and storytelling. He knows the power of story and the importance of tradition.

B. *Pauline*

Pauline, the second narrator, is female. She is a deeply religious Roman Catholic convert; more of an historian than Nanapush is; a writer, rather than a storyteller.

> ... I'll answer to the name I drew from Superior's hand. I prayed before I spread the scrap of paper in air. I asked for the grace to accept, to leave Pauline behind, to remember that my name, any name, was no more than a crumbling skin.
>
> *Leopolda.* I tried out the unfamiliar syllables. They fit. They cracked in my ears like a fist through ice. (205)

Nanapush is an important choice for this novel's first narrator because of the resonances that his name produces. The name of this second narrator is not so immediately appropriate. "Pauline" summons St. Paul the biblical apostle and writer, of course, with his epistles explaining rules and orders of conduct and doctrine to the early Christian churches; and "Sister Leopolda" recalls initially St. Leopold, who was, according to John Delaney, "known for his piety and charity." "Leopolda," however, is more than that. It is neither common nor as innocuous as it might seem.

First, there is a very real difference in the way that each narrator acquires her or his name. Nanapush's father gives him his name after much consideration, and with the knowledge that the name is significant, that it is indicative of character. It is given explicitly in the context of Chippewa tradition. Pauline, however, is known only by her "Christian" name until she is re-named by her Mother Superior. She appears not to have an "Indian" name, at all, as does Nanapush, who notes, "Nanapush is a name that loses power every time that it is written and stored in a government file. That is why I only gave it out once in all those years" (32). Pauline draws her new name from her Mother Superior's hand: it is written from the first, and it has been chosen for its appropriate aura of duty fulfilled. Curiously, the name actually is closer to an Indian name—it comes closer to matching the characteristics of the woman—than her original name ever was. In short, Pauline becomes in one sense close to being true to her heritage at just the moment when she repudiates that heritage for good. Pauline prays to be able to accept the name, and then to be able to live up to it.

Pauline's new name, "Sister Leopolda," is significant on several levels. According to the *New Catholic Encyclopedia*, the Leopoldine Society, for which she is surely named, was an Austrian mission society that was formed in 1828:

> Members of the organization were required to pray for the missions and to contribute a weekly alms to be used exclusively for America. Funds thus collected were ... distributed to needy bishops and religious communities in the U.S.... [Though] it ceased to exist in 1921.... this foreign mission aid proved more effective in strengthening Catholicism than the bare facts indicate, and served to arouse nativist opposition. (Blied 664)

Members of the organization included not only Bishop Baraga, who "worked with" the Chippewa in Marquette, but also, and perhaps more importantly for this novel, Baraga's colleague Francis Xavier Pierz:

> From 1835 to 1871 Pierz labored among the Indians of Michigan, Wisconsin, and Minnesota. He arrived there in 1852 and, from his headquarters at Crow Wing, traveled by foot and horseback to virtually every Chippewa village in the territory. As government policy restricted the Indians' domain, he worked to have their vacated lands settled by German and Slovene Catholics. Many Catholic communities in the

> Diocese of St. Cloud, Minn., owe their origin to his efforts. (Furlan 353)

"Sister Leopolda" thus brings together in one startlingly explicit and powerful image the history of Catholic missionary zeal among the Indians, specifically in Minnesota; the Indians' displacement off their land with the assistance of the missionaries; and the heritage, indeed, of Erdrich herself (as German-American-Chippewa stock). Pauline reflects on her new name, and decides that the "unfamiliar syllables ... fit." Indeed, they do.

C. *The Narrative*

The text is divided, then, between two contradictory narrators, one of whom addresses his story explicitly to a particular reader/listener, Lulu, a character who was first introduced in Erdrich's other fiction; and the other of whom is clearly going mad, as she descends into the non-referential morass of a religious zealot's mind. To increase the complexity, Nanapush claims to be delivering an oral history; he claims to be *telling* his story. Pauline, however, is clearly writing her story. She has no explicit audience like Lulu, for instance; in contrast to Nanapush, she is remarkably solitary, to the point of being anti-social. Thus a tension is also created between oral and written history in the tensions between the narrators.

Nanapush's narrative presents itself from the beginning explicitly as a traditional story, expressing itself in fictive terms nearly as sweeping and as general as "Once upon a time":

> We started dying before the snow, and like the snow, we continued to fall. It was surprising there were so many of us left to die. For those who survived the spotted sickness from the south, our long fight west to Nadouissioux land where we signed the treaty, and then a wind from the east, bringing exile in a storm of government papers, what descended from the north in 1912 seemed impossible.
>
> By then, we thought disaster must surely have spent its force, that disease must have claimed all of the Anishinabe that the earth could hold and bury.
>
> But the earth is limitless and so is luck and so were our people once. Granddaughter, you are the child of the invisible, the ones who disappeared when, along with the first bitter punishments of early winter, a new sickness swept down. (1–2)

On its first page, the novel establishes itself as story, and quite likely an oral story at that, because many of the connections which are vital in a successful oral tale are made in the opening sentence of the novel. We know from the chapter heading that the season is winter; winter in Chippewa country suggests snow; and "we" are connected personally to snow, and the history of snow, in the first sentence. The environment, the audience, the subject, the tale, and the teller are all skilfully and intimately connected, then, before the audience has even properly settled itself.

Next, connections are made between the "we" of the tale and nature: adopting a stance rather like the grand, mythic, sweeping approach taken by the writers of the Judeo-Christian Old Testament, Erdrich's narrator makes plagues appear, descending in this case on the Chippewa from the four points of the compass. The narrative here embarked upon is the story of what arrived in the winter of 1912, from the north, the fourth point of the compass: it announces itself as the story of the latest Chippewa plague, tuberculosis. Nanapush's story takes on a remarkable importance from the first. This is unsurprising; Vecsey claims that "in the main [the Chippewa] have relied on their storytelling for self-knowledge" (Vecsey, "Envision" 124); Copway agrees. He writes:

> The Ojibways have a great number of legends, stories, and historical tales, the relating and hearing of which, form a vast fund of winter evening instruction and amusement.
>
> There is not a lake or mountain that has not connected with it some story of delight or wonder, and nearly every beast and bird is the subject of the story-teller, being said to have transformed itself at some prior time into some mysterious formation—or men going to the stars, and of imaginary beings in the air, whose rushing passage roars in the distant whirlwinds. (95)

In addition to establishing itself as a Chippewa Winter's Tale, then, with all of the traditions that that implies, the narrative situates itself within the history of the people Nanapush calls the "Anishinabe": we are reminded of the sweeping death of the smallpox epidemic;[16] the steady displacement west of the Chippewa;[17] the "storm" of government treaties that were signed and that resulted in the Chippewa "exile" to reservations.[18]

The narrative on this first page does not forget to endow its narrator with authority; Nanapush establishes his credentials as authoritative story-teller at once, by naming himself as one of those who were "left to die," who had "survived the spotted sickness," who had taken part in the "long fight west," "signed the treaty," and had concluded vainly that "disaster must surely have spent its force." He sounds rather like Ishmael of Melville's *Moby Dick*, himself a reference to Job's servants in the bibical story, who, one after the other, come to Job with yet more evil news: "And I only am escaped alone to tell thee" (Melville 582; *Job* 1:15, 16, 17, 19).

But not only is this story and its grand style fitted for an oral presentation, it is suited to the kind of mythical reality that is so important to the kind of social system Nanapush represents, the kind of reality he encapsulates. Paula Gunn Allen writes: "The traditional tribal concept of time is of timelessness, as the concept of space is of multidimensionality. In the ceremonial worlds the tribes inhabit, time and space are mythic" (*The Sacred Hoop* 147).[19]

And Mircea Eliade writes:

> Basically, if viewed in its proper perspective, the life of archaic man (a life reduced to the repetition of archetypal acts, that is, to categories and not events, to the unceasing rehearsal of the same primordial myths), although it takes place in time, does not bear the burden of time, does not record time's irreversibility; in other words, completely ignores what is especially characteristic and decisive in a consciousness of time. Like the mystic, like the religious man in general, the primitive lives in a continual present. (86)

Finally, the beginning of the third paragraph on this first page has the ringing tone of a phrase used often and well, again rather like "once upon a time," or "they lived happily ever after": "But the earth is limitless and so is luck and so were our people once." It is the refrain of the story-teller, tempered by the lament of a representative of a people diminished by "disaster."

This kind of story-telling has a well-respected history among the Chippewa, where there is a strong traditional belief in the power and richness of Chippewa stories and their tellers. Coleman, et al. note:

> At present, the older Ojibwa seem to feel that they are the last link with a distinctly Indian past. They recognize the fact that those who can relate the native traditions with sureness and relative completeness are rapidly becoming few in number. Story telling as an art and a custom is now at time's edge among the Ojibwa, quite in contrast to the set rules formerly maintained and the high esteem in which the art was

> once held. It used to be true that some of the narratives were considered sacred and so could be told only by certain individuals. There was also a designated time for story telling and in this respect the Ojibwa, like many other North American Indians, regarded winter as the only appropriate season. When the blustery winds blew against the stiff branches of the trees and piled the snow high on the trails, then young and old gathered in the warmth of the wigwam fire to listen to Ojibwa lore. (4)

In other words, the first page of the narrative corresponds exactly to the requirements of context and protocol of traditional Chippewa story-telling. *Tracks*, at least this part of it, must be seen in this self-referential context if we are to make sense of it.

This is, then, according to the traditions summoned in the first three paragraphs, an oral story, yet it has begun by setting itself in another context: it not only has those novelistic conventions called chapters, but it is contained within the covers of an (obviously printed) novel. The narrative proclaims itself a Chippewa story, has a Chippewa setting, and uses a narrator whose name is prominent in Chippewa legend, yet there are English translations provided for the Chippewa words used, and the text itself is in English: this novel/story is obviously meant for an English-speaking audience, not a Chippewa-speaking one. *Tracks* as "told" by Nanapush, then, is an oral story, a story told for a reason, redolent with the rich history of storytelling that is implied in Chippewa history.

However, as I have already begun to indicate, this is also emphatically a written story, a novel, with the white and Judeo-Christian implications that that tradition implies; indeed, Pauline claims to be white at one point, just as she is becoming most delusional about her religion:

> Every day I saw more clearly and I marveled at what He showed me. For instance, exactly where I was from. One night of deepest cold He sat in the moonlight, on the stove, and looked down at me and smiled in the spill of His radiance and explained. He said that I was not whom I had supposed. I was an orphan and my parents had died in grace, and also, despite my deceptive features, I was not one speck of Indian but wholly white. He Himself had dark hair although His eyes were blue as bottleglass, so I believed. I wept. When He came off the stove, his breath was warm against my cheeks. He pressed the tears away and told me I was chosen to serve. (137)[20]

But it is not only when Pauline descends into her mad private world that her narrative becomes distinct from Nanapush's. She begins, in her first turn as narrator:

> The first time she drowned in the cold and glassy waters of Matchimanito, Fleur Pillager was only a child. Two men saw the boat tip, saw her struggle in the waves. They rowed over to the place where she went down, and jumped in. When they lifted her over the gunwales, she was cold to the touch and stiff, so they slapped her face, shook her by the heels, worked her arms and pounded her back until she coughed up lake water. She shivered all over like a dog, then took a breath. But it wasn't long afterward that those two men disappeared. The first wandered off and the other, Jean Hat, got himself run over by his own surveyor's cart.
>
> It went to show, the people said. It figured to them all right. By saving Fleur Pillager, those two had lost themselves. (10)

Pauline's narrative begins in the middle, or certainly at a dramatic moment in the narrative line. It spends no time establishing the credibility of the narrator or the history of the story; it is addressed to no one in particular; it is detached from its roots. This is shown, for instance, in the words Pauline uses to describe the village people who witness Fleur's drowning and "rescue." For Pauline, these are not "we" or "us." She does not side with them at all: they are, merely and impersonally, "the people"—those people—others. Pauline's narrative does not partake in the mythic reality of Nanapush's story, either. Her account falls nicely into Paula Gunn Allen's description:

> The way of the imagination is the way of continuity, circularity, completeness. The way of the intellect is the way of segmentation, discontinuity, linearity. ("Bringing Home the Fact" 563)

Where Nanapush stresses "continuity, circularity, completeness," Pauline stresses "segmentation, discontinuity, linearity."

The balancing of the two narratives is all accomplished with a distinctively playful postmodern sense of ironic detachment: the narrators accuse each other of inaccuracy, even call each other liars, from time to time: "[Nanapush] said this to me in the old language, and the words were strong and vulgar.... 'You filthy mouth,' I said, my tongue loose and unbridled. 'I hope the devil tears you apart piece by piece and fries each morsel!'" (147)

"[Pauline] was different once her mouth opened. She was worse than a Nanapush, in fact. For while I was careful with my known facts, she was given to improving truth" (39). Nanapush obviously thinks Pauline is an appropriate butt of ribald jokes; Pauline thinks Nanapush is a lewd, disgusting little man.

Tracks has a scrupulously accurate sense of socio-historic details, which are self-consciously and painstakingly constructed. It is important, for instance, that so little is made of the "world history" attached to the dates that are so scrupulously given. The fact that it is 1914–18 and Indians go to fight in WWI takes a distant second place to the importance of a sense of place and daily history: the novel has a postmodern sense of history as personal narrative, in fact. The logging and loggers are more important than the war. Story is more important than so-called "fact":

> I told this to Fleur that same day. I made her sit down and listen, just the way you are sitting now. Your mother always showed the proper respect to me. Even when I bored her, she made a good effort at pretending some interest. She never tapped her fingers on her uncovered knees, shuffled and twisted and made faces out the window like you. (178)

The growing rift between oral and written, Nanapush and Pauline, widens as the novel advances. Nanapush tells stories when Margaret has her head shaved, but cannot stop that disaster from happening; Fleur believes papers have "no bearing or sense" (174). It is Father Damien who signs papers (172), who smooths out maps (173), while Margaret "traced the print she could not read" (173), and manages to misinterpret. Pauline's narrative, however, while it partakes of an attitude typical of magical realism (Pauline's first words are concerned with the "first time" Fleur Pilager drowns, for example), is stable and written from the first.

Tracks is, in short, an English novel, self-reflexively posing as a Chippewa story, or else a Chippewa story, trapped inside the body of an English novel, or else both.

The playful gaming of the two also circles around the richness of the "tracks" of the title. Tracks are mentioned repeatedly throughout this novel. We hear of the tracks of the words across the page: "She swiped at the sheets with her hand, grazed the print, but never quite dared to flip it aside. This was not for any fear of me, however. She didn't want the tracks rubbing off on her skin. She never learned to read, and the mystery troubled her" (47).

We hear of the tracks of the Chippewa: "In my fist I had a lump of charcoal, with which I blackened my face. I placed my otter bag upon my chest, my rattle near. I began to sing slowly, calling on my helpers, until the words came from my mouth but were not mine, until the rattle started, the song sang itself, and there, in the deep bright drifts, I saw the tracks of Eli's snowshoes clearly" (101). And of more metaphysical tracks: "In the morning, before they washed in Matchimanito, they smelled like animals, wild and heady, and sometimes in the dusk their fingers left tracks like snails, glistening and wet" (72). And there are also animal tracks: "I think like animals, have perfect understanding for where they hide, and in my time I have tracked a deer back through time and brush and cleared field, to the place it was born" (40). Finally, of course, there are the intertextual tracks of characters through this novel from the other novels in the cycle.

Worlds and philosophies and novels collide in many ways in *Tracks*: there is a clash of cultures, for instance, when Indian "medicine" and religion, and customs of marriage and burial are opposed to the Catholic Church, and its teachings:

> By then, I'd traded candles and ribbons for the thing I needed from Moses, who made the dreamcatcher. He gave me the sack of medicine powder, then held my eyes with his and made me tell him whom I meant to snare. He dragged Eli's name from me in a whisper, which caused him great amusement, and then his face twisted.... but already it was too late. I couldn't stop myself. The dust Moses had concocted was crushed fine of certain roots, crane's bill, something else, and slivers of Sophie's fingernails. I would bake it all in Eli's lunch.
>
> From the first day, I knew what I planned was possible. (80)

> There are two plants. One is yarrow and the other I will not name. These are the sources of my medicine, and I used them for the second time on Fleur, the third time on a Pillager. Only because of Pauline, I did not complete the job. (188)

> By the time we got our foreheads crossed with ashes, we were keeping company. "I hear you're thinking of exchanging the vows," said Father Damien as we shook his hand on our way from the church. "I'm having relations with Margaret already," I whispered to startle him. "That's the way we do things." (123)

"Medicine" and other Indian customs are opposed to Father Damien's belief in confession, or his urging of his congregation to get married. The explicit racism of the Catholic Church is masked to some extent by Pauline's madness, but it is there, for instance, in the decision to refuse to allow Indian girls to be nuns: "'The Indians,' I said now, 'them.' Never neenawind[21] or us. And I soon found it was good that I did. For one day during supper Sister Anne announced that Superior had received word that our order would admit no Indian girls..." (138).

The rift between cultures is also echoed in Fleur's confidence that no one will want to live on her land, that they will respect history and Pillager tradition:

> "They know better," Fleur said with confidence that seemed pitiful and false to me, though I had never before pitied Pillagers. "They won't dare to throw us off the shores of this lake," she promised. Father Damien had already heard otherwise. "There's some who want to build a fishing lodge," he said in a gentle voice. "They're willing to trade for an allotment someplace else." Fleur refused to hear this, but I could not ignore it and digested this new betrayal in silence. (175)

In all of this, one must note two things: first, although the novel is not necessarily "fun" to read, it is very playful. It is not a sombre tract on the evil of encroaching civilization or a Romantic declamation on the "noble savage." Second, because of all the flitting back and forth that goes on between camps in the novel, the reader is forced into an active position. One is forced to play between the borders that are set up as the limits of the novel, to have an active knowledge of enough Indian history and mythology to know what Nanapush is about when he claims the attributes of his namesake, yet is not forthcoming enough to supply all useful information about himself. The novel risks being misunderstood completely because it assumes that its readers will be educated, or will become educated enough, to know about the windigo, to know about Nanapush, to know about the Dawes act and the allotment laws and the exploitation of the Indians by the Whites who took over their land. Nanapush describes his novel accurately when he muses: "Snares. They demand clever fingers and the ability to think exactly like your prey. That requires imagination..." (118).

3. Conclusion

Into this complex playground of shifting narrators and shifty story-lines, with its blend of real and mythological characters, magic and history and storytelling, Erdrich places her readers: one inside the novel, in fact, as reader and participant. Lulu is one who can read here her own history, who is apparently tracing her own history and her own story, her own narrative, within and around the narrative of Nanapush, rather than following it painstakingly through the vagaries of western history books, which will tell her little of what she wants to know.

Lulu, then, is not exactly a reader, but a listener, Nanapush's granddaughter and audience, a character with a history that is told here. She is borrowed from Erdrich's other fiction in what has become a quartet of novels, fiction which was written first, but set later. One is required to know her past, which is her future, and put the two together, and bring the story up to date. The other reader is Pauline's audience, and is not a listener or a character, but very much a reader.

In any case, the interpretation of this novel is not and cannot be fixed, but an interpretation of some sort is required; this is a profoundly gameful fiction, with limits and demands, and not "just" play. Hesitating between narrators, hovering between worlds, the novel requires that the reader do the work involved to create a novel, not just consume one, and not just dance around one.

The narrative of *Tracks* hesitates between two worlds. The reader is asked to choose between and among them, or to balance them, as need be. In effect, the reader is asked to play a game, balancing the playful ribaldry of Nanapush with the sombre religiosity of Pauline, the fluid movement of the Chippewa religions and oral storytelling, with the rigid structure of Roman Catholicism. There are (at least) two worlds here, but not two worlds that clash, exactly, so much as two worlds that sparkle with an intense kind of playful interactive rivalry.

CHAPTER SIX

Conclusion: Spinning Playfully into the Future

There is no simple way to say this.

(Michael Joyce, *Afternoon, A Story*)

There are always endings, but there are not always conclusions.

(Coover, *Pinocchio in Venice* 45)

Standing in the National Gallery of Canada in Ottawa, Ontario, at the foot of what must rank as the single most controversial painting ever acquired by that gallery, Barnett Newman's *Voice of Fire*, one is struck by several things at once. First is the certain knowledge of controversy, the awareness that people all over the country were upset, or even angered, by the fact that the Gallery would buy the painting hanging there on the wall. Second, and vastly more important and impressive, is the sheer amazing presence of the piece.

The controversy that swirled around *Voice of Fire* was double: first was the cost of the painting; second was its apparent simplicity and palpable "lack of content."[1] *Voice of Fire* cost the National Gallery $1,758,000. The gallery claimed that that was a bargain; "the public," as represented in "the media," perceived the purchase as a waste of money. As for the "content" of the painting, Harold Rosenberg writes: "A Newman such as ... *Voice of Fire* ... puts it squarely to the spectator to discover within himself whether religious or heroic associations are any longer needed in order to experience awe and tranquility" (32). Newman's paintings, at least the ones in the group that includes *Voice of Fire*, aim at creating "a sense of epiphany without concrete associations" (38); in other words, they are precisely, and pointedly, *about* a "lack of content."

The outcry against *Voice of Fire* indicated many things, but indifference was not among them. The "general public," whoever that is, is apparently centrally concerned with Art, or at least Art as National Gallery acquisition, Art as commodity. Systems of patronage that have traditionally supported the arts are in decline; large corporations like Molson's are often more interested in large and uncontroversial sports events and spectacles at which they can promote themselves, than in art; money for art is increasingly expected to come either from sales (often to large institutions like the National Gallery, and, therefore, from "taxpayers" indirectly), or from governments directly (and therefore from "taxpayers" directly). In addition, governments have been telling taxpayers for some years now that governments generally are out of money. Because of the lack of government funding, services intended directly for "the general public" have been cut back. The "general public" is concerned with art, then, in part because it sees itself as increasingly responsible (or at least as being held to be more responsible) for paying for what is called art. If that art seems centrally concerned with mocking them, or performing opaque intellectual feats that they don't understand, or even with ignoring them altogether, then it is no wonder that they have become testy about supporting its creators.

This is, in part, a simplified version of what Fredric Jameson, Charles Newman, and others have called the "commodification of art"; this, they have claimed, is one of the markers of postmodernism, or "late capitalism." Perhaps more remarkable than the notion of art as commodity (it has always been that, really: a patron needed to approve of a piece, or at least not to be offended by a piece, or he might on a whim cut off an artist's funding or body parts), however, is the notion that "the public" should have a voice in what is purchased by the National Gallery. The "democratization" of art and artistic taste has gone some way toward removing the aura of inaccessible elitism of which postmodernism has accused modernism; perhaps to its credit, and perhaps at a cost of certain kinds of innovation, postmodernism is not only part of the "high art" world, but a part of "popular" art, as well.

The second major controversy about *Voice of Fire* was its lack of content: the painting consists of three vertical stripes, a centre red one flanked by two blue ones. It is, after all, not obviously "about" anything. What is more, it does not obviously "represent" anything. What is still more, it does not appear (at least in the grim and grainy black and white photographs that appeared in newspapers) to have been particularly difficult technically to make. As the protests reached their peak, Felix Holtmann, MP for Portage-Interlake, Manitoba, rose to prominence in the debate through his "endlessly repeated mocking statement—that he could produce a similar painting with a couple of cans of paint, two rollers, and 10 minutes of effort" (Yaffe 874).

But what is immediately impressive about *Voice of Fire* is not its "content" or the critical narratives that have been built up around it, but its sheer scale. Viewed in person, *Voice of Fire* is breathtaking. It must have been an extraordinary feat to create it. What it "says," what it is "about," has nothing whatever to do with its content, or with what it "represents"; what is important is the response it provokes. It is not "important" for its content, nor its technique, nor its thematic or technical innovation. What makes it impressive is not any of the qualities that are usually discussed in traditional art criticism, in fact. There is, instead, a strong sense in which the painting does not exist unless it is viewed—experienced—in person; it has, after all, no content that can be paraphrased. The point is this: it was designed precisely, and only, to provoke a response in a viewer. The art "works" only in the viewer's experience of it.

Thus in order for a viewer to "get" this piece of art, that viewer must do two things: first, she must be willing to set her preconceptions free, and play with what she has been given. Second, she must adjust her parameters to include a lack of content, a lack of representation, as a part of her potential experience of the painting. *Voice of Fire* sets up certain parameters within which to enact its game. It challenges preconceptions, and in so doing provokes a remarkable experience.

Postmodernist literature is likewise often seen as shallow, repetitive, empty, easy. It has been called a "Literature of Exhaustion," a literature that does not "say" anything. What I have argued in this study is that part of the apparent "depthlessness" and relentless playfulness of postmodernist literature lies in the techniques and preoccupations of postmodernism itself. Those techniques and preoccupations are perhaps best suited to expressions other than traditional literary ones: the rapid-fire and disorienting action sequences of a music video on television or *Natural Born Killers*, for example.

I have also maintained that in its literary manifestations, postmodernism needs the limits, rules, and borders that literariness supplies: the postmodernist novel, after all, participates in the archive known as "novel" as much as it does in the archive known as "postmodernism." My argument maintains, then, that the postmodernist novel plays, that it lacks final closure, certainty, or a precise meaning; but that it also necessarily needs limits to make its play possible, and to make it meaningful, or at least to make meaning possible. In that, the postmodernist novel constitutes itself not only as a participatory experience, but as game-like in some important senses. That it is, or can be, fragmented, discordant, disconnected, and without coherent plots or characters or settings or themes[2] is part of its point, as much as to be without an obvious representational referent is part of the point of *Voice of Fire*.

I have argued that literary postmodernism is a gamelike phenomenon in that it conforms to the stipulations regarding gamefulness in its twin obsessions with play, freedom, lack of constraint; *and* with rules, limits, borders. I have demonstrated that literary postmodernism expresses itself in the intertextual literary playfulness of Barth's *The Last Voyage of Somebody the Sailor*, which displays its boundaries as the ones described by western literature, in autobiography, in so-called "realistic" modes; as well as in the limits described by the exotic worlds of the east and the stories of Sindbad. Barth situates his novel in a world—an ocean—of story, and gives his novel boundaries that both expand and contract the edges of his novel in alarming and delightful ways. Barth's are boundaries that are based on the texts he chooses. I have also discussed literary postmodernism in the playfully dancing worlds of Erdrich's *Tracks*. There, storytelling is a way of life and knowing, a profound way to confront historical reality, land claims, interracial differences. Thus I have discussed literary postmodernism within two of the favoured archives of literacy: literature and history and their limits and borders.

There are other models, other writers, other possibilities altogether that are related to and conform with the suggestions of gamefulness and postmodernism that could have been, and that could be, explored in other novels, and in other spaces.

For instance, "hyperfiction" (or "hypertextual fiction," or, more simply in many literary circles, "hypertext") is often raised in discussions of postmodernism. Robert Coover (with George Landow, among others) teaches a course in experimental hyperfiction writing at Brown University. Coover's novels and stories reflect his interest in electronic writing—and have done so for some time. George Landow calls "The Babysitter" a hypertext story on paper, pointing to its multiple story lines, open choices, narrative possibilities and its general undecidability (*Hypertext* 108) as evidence of its hypertextuality.

Coover has this to say about hypertext: "With hypertext we focus, both as writers and as readers, on structure as much as on prose, for we are made aware suddenly of the shapes of narratives that are often hidden in print stories. The most radical new element that comes to the fore in hypertext is the system of multidirectional and often labyrinthine linkages we are invited or obliged to create" ("The End of Books" 24). This recalls John Hawkes's dictum about plot, characters, setting, and theme giving way to structure in fiction; it also draws attention to the concerns I have been raising in this study. Landow has argued that hypertext is the embodiment of post-structuralist conceptions of textuality, that hypertext is the practice that proves those theories. And I have argued that literary postmodernism continues to structure the elements of play according to the strictures of literariness, according to the demands of the novel, and of the book.

One of the more widely respected and quoted books on hypertext was written by Jay David Bolter. Bolter has noted this tension; he writes that "In all modern fiction, there is a tension between the linear experience of reading and the structure of allusion and reference..." (135).

Furthermore,

> The conceptual space of a printed book is one in which writing is stable, monumental, and controlled exclusively by the author. It is the space defined by perfect printed volumes that exist in thousands of identical copies. The conceptual space of electronic writing, on the other hand, is characterized by fluidity and an interactive relationship between writer and reader. These different conceptual spaces foster different styles and genres of writing and different theories of literature. (11)
>
> Electronic writing emphasizes the impermanence and changeability of text, and it tends to reduce the distance between author and reader by turning the reader into an author. The computer is restructuring our current economy of writing. It is changing the cultural status of writing as well as the method of producing books. It is changing the relationship of the author to the text and of both author and text to the reader. (2–3)

And as Landow points out, the tensions between author and reader, text and writing, mutability and permanence are resolved—or disappear— when post-structuralist theories are put into practice in hypertext, with its shifting, unstable notions of what constitutes a text, and with its questioning of what activities actually constitute reading:

> One major effect of ... nonsequential reading, the weakening of the boundaries of the text, can be thought of either as correcting the artificial isolation of a text from its contexts or as violating one of the chief qualities of the book. (*Hypertext* 62)

Coover has not, to date, published a volume of his own hypertext writings, but his work now, even more than it has in the past, reflects a playful concern for structure, bounds, and limits. His recent novel *Pinocchio in Venice* describes an excursion into a city that is famous as a labyrinthine structure (it is called in the novel a "dreary Venetian labyrinth" [250]), that is renowned for its extravagant carnivals, and its attendant play with conventions and boundaries. The excursion is taken by a grown-up Pinocchio (past yearning to become a "real boy," he is now a humanities professor) working on his last book: he is finishing his autobiography. The terms in which Professor Pinocchio describes his book and his purpose in Venice are also the terms he uses to describe Venice itself:

> [There is] a kind of itchy boundary between everywhere and somewhere, between simultaneity and history, process and stasis, geometry and optics, extension and unity, velocity and object, between product and art. (20)

Pinocchio has gone to Venice to find a metaphor that will sum up his life and his work; he has gone there in search of literature, only to find a labyrinth, a watery maze that is only bridges and blind alleys.[3]

Pinocchio in Venice is full of speculations about writing, and autobiography, and the intersection between those and scholarship and fiction. It is important, with regard to Coover and hypertext, to note that Pinocchio's book is not a manuscript in any kind of traditional sense. He is not carrying a sheaf of papers when he arrives in the city, but a computer. In fact, the metaphors connected to writing and books come naturally to his mind, only to be rejected as essentially quaint and old-fashioned:

> Like a crisp clean sheet of paper, he thinks, and he is struck at the same time by the poignancy of this metaphor from the old days. For paper is no longer a debased surrogate for the stone tablets of old upon which one hammered out imperishable truths, but rather a ceaseless flow, fluttering through the printer like time itself, a medium for truth's restless fluidity, as flesh is for the spirit, and endlessly recyclable. (31)

His book is described, not as a manuscript at all, but as "his current work-on-hard-disk" (36).

And Pinocchio himself, of course, has been for some time a metaphor for what it is to be human:

> ... he renounced vagabondage and rebellion and idle amusements, and so, through discipline, has acquired that dignity which, as all the world insists, is the innate good and craving of every moral being; it could even be said that his entire development has been a conscious undeviating progression away from the embarrassments of idleness and anarchy, not to mention a few indelicate pratfalls, and toward dignity. Indeed, he is one of the great living exemplars of this universal experience, this passage, as it were, from nature to civilization—from the raw to the cooked, as one young wag has put it—or, as he himself has described it in his current work-on-hard-disk in the chapter "The Voice in the Would-Pile," "from wood to will." (36)

The playfulness of Coover's novel is to be found everywhere in it, not least in the extravagant puns and word-plays of the characters, the fractured quotations, the mangled references—of which it is full (see, for instance, "The Voice in the Would-Pile"). The limits of the novel, too, are describable, although, as befits the work of someone who is interested in hypertext, those limits are strange and shifty.

Coover's novel is bounded, first, by suggestions of hypertext and the limits of the book; second, by its use of Venice, as a labyrinthine example of a literary, aesthetic, artistic site and symbol; third, by labyrinths themselves.[4] And finally, there are obvious and unavoidable intertextual references to the Disney film of *Pinocchio*[5] (and, of course, to the Collodi story on which both novel and film are based[6]).

Coover's novel is not, of course, the only work that could be cited in these contexts. Literary postmodernism manifests itself repeatedly in terms of play and its limits. John Hawkes's novel *Whistlejacket*, for instance, bounds itself with an intricate plot involving murder and horses, and thus summons as an intertextual reference Arthur Conan Doyle's "Silver Blaze." *Whistlejacket* features the boundary-crossing artistic-medical pioneer George Stubbs as a literary character; it cites characters, instances, plots, and attitudes from Cortázar's story "Blow Up," and the Antonioni film of the same name; it uses fox hunts as a boundary marker[7]; and it discusses painting, photography, and literature, and the places where they intersect.

Hawkes's novel takes up a playful stance on language and literary conventions, and then plays with the places where that play must end, with the limits of the form.

Still other examples abound: Donald Barthelme uses as a limit the fairy tale of *Snow White*; Angela Carter repeatedly employs fairy tales in the same way; Thomas Pynchon's *Vineland* makes playful critical use of television; and still other examples suggest themselves, and multiply and divide themselves, spinning playfully into the future.

This, then, will not have been a conclusion, a bringing together, a closing up; but an opening out, a suggestion for other directions, other modes, other possibilities, still other boundaries—an ending, merely.

NOTES

Chapter 1

1. "The Player Played" has very recently been published as a scholarly monograph by Garland Press under the title *Theories of Play and Postmodernism*. My references will, however, be to the dissertation rather than to the book.

2. I examine this problem—that is, the problem of the relationships between games and novels—at more length in my article "Playing the Games we Read."

3. There is a kind of self-absorbed chauvinism associated with a good number of studies of American literature (and associated with that literature itself), and virtually by definition this study does not escape that chauvinism completely. By "society" in this context, I mean late twentieth-century North American society, which, admittedly (and sometimes explicitly, as in the case of some of John Barth's later work, for instance), is often by default assumed to be male, largely white, and predominantly heterosexual. I do not wish to imply that any of those positions are norms or givens; on the contrary, one of the strong lessons that postmodernism teaches is that genders, races, sexual orientations, and social classes are always constructions, and always at least potentially problematic.

As for games, there are many societies that are obsessed with games, but not necessarily with the same games in all cases, and not necessarily in the same ways that North Americans are. My use of baseball examples in this study limits some of my later discussions of games to a certain kind of audience; a more wide-ranging example (but a less "American" one—and this is a study of American literature, after all) might have been soccer. I have tried to vary my comments and examples to include a range of sports and games, but literature and the criticism and theory around that literature returns repeatedly to baseball. In other words, for all its limitations, baseball stands as a "literary" sport and game. Baseball may require some contextualisation for some readers, but this is evidence of the contextual nature of games. An interesting example of contextualisation is to be found in the explanatory baseball notes included as an appendix to the British edition of Coover's *The Universal Baseball Association*.

Salman Rushdie is eloquent witness to the possible dangers of playing certain games; a new generation of young Russian postmodernists (and writers of many other countries) also routinely experience profound consequences of play that the "we" of my remark does not address at all. In the context of this study, I should make this note: on one hand, Louise Erdrich is a political writer who is very aware of some of the difficulties of living on boundaries and of playing certain games; on the other hand, that there is a political stance implied, but rarely stated explicitly in some of Barth's work, for instance, may speak to both sides of some of the questions of chauvinism I mentioned at the beginning of this note.

Chapter 2

1. I leave for elsewhere a discussion of the extent to which television shows (including, but not only, those unfortunately named but boundary-crossing genres known as the "docu-drama" or the "info-mercial") are the real short stories of late twentieth-century North America.

2. *Forrest Gump* is notable for several reasons. First, it is a technical tour-de-force, placing Gump, as it does, into historical situations, with historical characters, using archival footage as seamlessly and convincingly as Woody Allen managed in his placements of Zelig. Second, it is extremely popular: this is a factor worth recalling in the context, for instance, of Umberto Eco's *The Name of the Rose*'s being available—and being a best-seller—in paperback in the racks beside grocery store line-ups. (On the importance of popular culture in postmodernism, see Lawrence Grossberg's "Putting the Pop Back into Postmodernism." Grossberg writes that "specific postmodern practices are often most powerfully present precisely within

forms of popular culture" [177], like movies and television shows.) Third, it is being widely read and appreciated—even celebrated—as a parable about the innate worth of the "common man." Reviewers have gone so far as to label the film a tribute to stupidity; indeed, it is there that the money-making possibilities of the film stretch as widely as Gump's own abilities.

That "popular" reading often obscures a more sophisticated reading of *Gump* as a parable and allegory of America itself in a postmodern age. *Gump*'s "common man" hero can be read readily as deconstructing part of America's myth about itself, with Gump himself, after all, not a "common man" hero so much as just a profoundly simple-minded man (see the *OED* definition of a "gump"), really competent in only one activity (running), as manifested in his football career (probably nothing is more quintessentially American than a football hero: O.J. Simpson is spectacular vindication of that view). Mixed with his singular running ability, of course, is an uncanny knack for making money.

Whether simple-minded hero or merely simple-minded, however, Gump is in the middle of most of the popular cultural fads of the 'fifties, 'sixties, and 'seventies, from "mooning" to "happy face" tee-shirts, from "Shit Happens" bumper stickers to being the country's pre-eminent running guru. Gump is even responsible for Elvis Presley's famous hip-swinging dance (of his entire act, that was probably the single element that made Presley such a commercial success). And Gump appears in less innocuous situations as well: he goes to Vietnam; he is present at the Watergate break-in; and his great love, Jenny, dies near the end of the film of an unnamed virus that is likely to be AIDS.

My most significant interests in the film, however, are these: first, its frame story, setting; and metafictional concerns. The film is a transcription of Gump's life story, told by him while sitting on a park bench. This situates it in a long tradition of oral tall tales (see, of course, Twain's "The Celebrated Jumping Frog of Calaveras County") and story-telling more generally (he is even accused once of making up his story, and then reassured that he does tell it well, in any case). Gump's exploits are usually televised, and he is cheered on throughout the film by a kind of chorus of three men who seem to do nothing but watch television. These features, combined with constant intertextual and historical references (from Richard Nixon to Abby Hoffman, the Black Panthers to Dick Cavett, John Lennon to "Bear" Bryant), make *Gump* from the beginning self-reflexively metafictional.

Second, *Gump* is a parable about the general and growing importance of the electronic media in American culture. Not only are the wonderful, self-reflexive technical tricks that are used in the film important, but television itself is ubiquitous: Gump meets Elvis face-to-face, but only actually sees Elvis perform his dance on television; and personal contact has disappeared by the time of Gump's later adventures, which are television spectacles entirely (the shooting of Reagan, for instance, is a passing television news story, only one of a string of casual shootings that help hold the film together structurally).

Third, all of Gump's enterprises are profoundly commercial: everything he touches makes him money. This becomes the true value of those various enterprises, in fact (see the invention of the "Shit Happens" bumper sticker or the "Have a Nice Day" happy face: both of those products, Gump notes, made their inventors "a lot" of money).

Finally, the film self-consciously pokes fun at one of the icons of American popular myth and culture: the rugged individual and his living of the American dream. Gump himself is, above all else, a deliberately simple-minded parody of the much-vaunted American individualist who succeeds in part by ignoring the problems that rage everywhere around him. His ancestor was one of the founders of the KKK; his mother is forced to have sex with his school's principal in order to get Gump enroled; with an IQ of 75, Gump actually graduates from college, which he attended on full scholarship; Jenny's dream of being a singer materialize only when she appears on stage in a strip club (as "Bobbi Dylan," singing "Blowin' in the Wind"), wearing nothing but a guitar; even the braces that Gump wears at the beginning of the film are designed specifically to straighten his back, to make him less "crooked than a politician." Gump's successes begin, significantly, when he breaks free of the cumbersome braces and begins to run. As the braces break away, Gump barges through and past a Southern chain gang.

Read unironically, *Gump* is a tract that argues for the realistic possibility that the so-called American dream can come true for anyone; read ironically, it is a profound commentary on the mindless vacuousness of at least certain aspects of American capitalism. When read ironically in combination with its technical brilliance, its metafictional nature, and its popularity, this more sophisticated reading turns *Gump* into postmodernism: it is sentimental nostalgia for the 'sixties and 'seventies, and can be read successfully that way (as *The Name of the Rose* can be read as a detective story...), but it is also a technically brilliant avant-garde allegory that deconstructs the "American Dream" while at the same time living it by "making a *lot* of money." In this second reading, America is Gump, bumbling and simple-minded, if apparently well-meaning; he is moral on an individual level (but still unquestioningly willing to go to Vietnam, on one hand, and appallingly ignorant of numerous very real social problems, on the other); he is cheerfully good-hearted and generous, sharing selflessly stories and chocolates; but finally, he is opportunistic, exploitative, and capitalistic, above all.

And so Gump sits, telling his story on a park bench, apparently believing that life really is "like a box of chocolates": you may really never know what you're going to get, but if you're a simple-minded all-American "good guy" like Gump, at least you know that whatever you get, it will be sweet. Straightforwardly, this is a feel-good message of hope, but this film begs to be read ironically; it begs to be seen as postmodernism, as deconstructing itself and the very myths that it depends upon for its success.

For a more extended discussion of *Forrest Gump* as postmodernism, see my "Like a Box of Chocolates."

3. Whereas a naive reading of *Forrest Gump* constitutes that film as a feel-good romance, a naive reading of *Natural Born Killers* will most likely render it a well-made, though essentially "B-grade" action-adventure film. (This reading will be less common for viewers of the "Director's Cut" of this film, in part because Olive Stone discusses the film explicitly as a dark social satire.)

Like *Gump*, *NBK* is certainly aware of its own technical possibilities and limitations: it employs the self-reflexive filmic techniques most often associated with television music videos, and with the attendant rapid cuts and mixing of images, intertextual references, and camera angles from that genre, it creates what is possibly the most hyperactive movie ever made, a film that surges over the viewer with a barrage of image and sound that is truly formidable.

Images, especially images of violence, have displaced language as the semiotic markers of note in this film. The importance of image is established first of all because television is ubiquitous. Television turns Mickey and Mallory into heroes in the first place. They look good when they are photographed, and so they can be transformed into profitable and popular television images, regardless (more cynically, because) of their actions; for instance, they star, with spectacular success, on a television show that features a "serial killer of the week." (In one of the more memorable lines from the film, Mickey notes that being second in a ratings race behind Charles Manson is nothing, really, to be ashamed of: "You can't compete with the King.") The domination of television in this world is underlined in a scene strongly and ironically reminiscent of Peter Sellers's role in *Being There*: Mickey tries to use the television remote control to change what is "on." Whereas Chance tried and failed to change the channel on an unpleasant "real-life" experience in *Being There*, in *NBK*, the scene outside the motel room window changes with the remote as though it were a screen, but Mickey has no control over what is on the television itself. Image is such an important marker, in fact, that it is used to separate levels of morality in this film: the two killers in the movie, though despicable in many ways, at least have a sense of values; killing, they claim, is pure and uncluttered, and accordingly, Mickey and Mallory see the world (literally) in black and white. And one of the more important instances of the image-based mixing of techniques and genres in *NBK* occurs during a particularly self-consciously staged jail break. The super-human and frankly unrealistic escape exploits of Mickey become cartoon-like, and the images on the screen promptly change into animated ones. The technique of mixing live action and animation, of course, is not unique to this film (see *Who Framed Roger Rabbit?* for example); but the technique in this movie is central, and not merely clever; the film is precisely *about* a media that can remake Mickey and Mallory into animated, and therefore value-free and attractive hero-images, instead of the murderous thugs that they are.

Image has displaced (or at least is indistinguishable from) language; violence has displaced productive ability; television has displaced reality. Mallory's family, at the beginning of the film, is explicitly a nineteen-fifties-style American network television "sit-com" family: that family image is itself a popular, pervasive, and nostalgic cultural fiction, seen here from an alternative angle. Ozzie and Harriet are deconstructed, in short; Mallory's home is far from happy, but its unquestioned patriarchal structure is instead a haven for violence and incest. And the film ends with yet another myth from American popular culture, as Mickey and Mallory, now retired, are pictured heading down the highway in their motor home, accompanied by *their* family and their family pet. The 1950s nuclear family has been displaced by the 1990s one: one image of America has displaced an older one. It is significant, finally, in an allegorical reading of this film that places Mickey and Mallory as themselves images of America, that the one act that

Mickey and Mallory regret in their life of murder and mayhem is the killing of an Indian whose only crime was to welcome them into his home.

If image and television are ubiquitous in this film, they are no more so than the casual violence that so permeates it. Mickey and Mallory are born out of a world dominated by, and in fact defined by, the media generally (and television in particular), on one hand, and violence on the other: Mallory's family is a corrupted television image; and Mickey makes history of his ancestry: "You call me violent," he claims at one point, "Well, my daddy was violent before me, and his daddy was violent before him: I was born into violence."

Mickey and Mallory, "natural born killers," media creations, media darlings, are remarkably attractive and repulsive at the same time. In addition, and disturbingly, they are the only characters in the movie with any sense of morals at all, though those morals are skewed. *NBK*, finally, is about America, and American reliance on electronic media, especially television; it is about the remarkable power that the media wield; it is about a lack of values beyond weekly ratings; it is about our love affair with media violence, including our willingness to be entertained by Mickey and Mallory, not only "in" the film, but also in the actual film; and it is about our willingness to make anyone who looks good on camera into a hero, or at least the image of a hero. *Natural Born Killers* is a disturbing and effective postmodernist film about images, finally, and their use, and our obsession with them. One of the more disturbing aspects of this film is the degree to which it works.

4. Stanley Elkin even uses the term to describe unconventional tuxedos ("tuxedo" itself, incidentally, is a term invented to describe then-unconventional formal men's wear): "Some of the younger guys wear post-modern tuxedos. Blair Underwood, a lawyer for the home firm on *L.A. Law*, has dark sequins hanging down the arms of his tuxedo jacket like a kind of glazed hair. Several fellows wear black running shoes with their tuxedos—formal Reeboks, dress Nikes. I see, no shit, a leather tuxedo. And there's another man in a tux with a long rabbinical coat over it. And another whose bow tie spills over his shirtfront like a growth" (52).

5. Foucault defines archives in this way: "Between the *language* (*langue*) that defines the system of constructing possible sentences, and the *corpus* that passively collects the words that are spoken, the *archive* defines

a particular level: that of a practice that causes a multiplicity of statements to emerge as so many regular events, as so many things to be dealt with and manipulated. It does not have the weight of tradition; and it does not constitute the library of all libraries, outside time and place; nor is it the welcoming oblivion that opens up to all new speech the operational field of its freedom; between tradition and oblivion, it reveals the rules of a practice that enables statements both to survive and to undergo regular modification" (*The Archaeology of Knowledge* 130).

6. This tendency shows up in even a superficial survey of titles, across specialties and disciplines; see, for instance, the influential "journal of postmodern literature," *boundary 2*; or Iain Chambers's *Border Dialogues*, in part a discussion of the ways "the traces of the past ... flare up in the present," and the way the "experience of the baroque [for instance] is also a sign of our times" (1); or Albert Borgmann's *Crossing the Postmodern Divide*, a philosophical attempt to "chart the passage from the present to the future," to investigate "the close not only of a century and a millennium but of an era" (2); or Bill Nichols's *Blurred Boundaries*, an investigation of "One of the most blurred of recent boundaries ... [that] between fiction and nonfiction" (x), which takes as one of its leading displays the widely televised trials involving Rodney King; or Robert Holub's *Crossing Borders*, an investigation into theories of literary reception and the fuzzy line between maker and consumer; or, of course, Derrida's "Living On: Border Lines"; there are many others.

7. Jameson thus situates his hypothesis squarely in the context of Paul de Man's warning that "the 'postmodern approach' seems a somewhat naively historical approach" (*Resistance* 119–20); Jameson's project becomes in part, then, an attempt to re-historicize without being naive. Foucaultian anti-linear notions of history as archive and archaeology thus become important for Jameson; his terms are "cognitive maps" and "genealogies."

8. Jameson makes this point in many places: in addition to the comments in *Postmodernism*, for instance, see his "conversation" with Anders Stephanson: "All in all, these developments have to be confronted as a historical situation rather than as something to be morally deplored or simply celebrated" (12).

9. There is a strong sense in which we are not suffused with meaning at all, but are, rather, suffused, or even gorged, with information; as in advanced autism, however, the meaning seems to have "dropped out" of our communication (see Donna Williams's *Nobody Nowhere*, and her interviews with Peter Gzowski on the CBC Radio program "Morningside" conducted in 1993 and 1994).

10. See McHale's *Postmodernist Fiction*, p.7, and n.13, p.237. Lodge's repertoire includes six items, not the five listed by McHale: Lodge lists "permutation" in addition to McHale's five "strategies" of contradiction, discontinuity, randomness, excess, and short circuit (see Lodge's *Modes of Modern Writing* 220–45). McHale next cites Ihab Hassan's "seven modernist rubrics (urbanism, technologism, dehumanization, primitivism, eroticism, antinomianism, experimentalism), indicating how postmodernist aesthetics modifies or extends each of them" (7). That list is from Hassan's 1971 essay "POSTmodernISM." McHale might better have made use of a more recent list; Hassan has been busy expanding what he himself calls a table of "schematic differences" between modernism and postmodernism, admitting that that table is only a place to start a discussion of differences. Hassan's more recent list of oppositions has thirty-five entries, some of them doubled or trebled (e.g., "Art Object/Finished Work" vs. "Process/ Performance/ Happening"): see "Toward a Concept of Postmodernism" 152. McHale has also transformed Wollen's scheme into six "oppositions," when Wollen lists seven. Wollen himself lists these: narrative transitivity vs. intransitivity; identification vs. estrangement; transparency vs. foregrounding; single vs. multiple diegesis; closure vs. aperture; pleasure vs. unpleasure; fiction vs. reality (see *Readings and Writings* 79–91). Finally, McHale cites Fokkema's "compositional and semantic conventions of the period code of postmodernism (such as inclusiveness, deliberate indiscriminateness, non-selection or quasi-nonselection, logical impossibility)," and notes that they "[are contrasted] generally with the conventions of the modernist code" (7; see Fokkema's *Literary History, Modernism, and Postmodernism*).

11. Thiher is, of course, aware of the complex relationship between past and present in postmodernism. See, for instance, his discussion of Fuentes's novel *The Death of Artemio Cruz*: "The autobiographical project is circumscribed by the limits of the pronoun that grammatically as well as

existentially—and one adverb implies the other in Fuentes—cannot transcend a present discourse rooted in the 'nowness' of the body. Time past cannot enter into this project, and the 'I' voice can only speak self-reflexively of its own limitations" (141). Perhaps Thiher's own discourse is "rooted in the 'nowness' [that it] cannot transcend"; perhaps Thiher's work needs to be read itself as a postmodernist and a-historical document.

12. That project has been attempted in several places; for one of the more engaging versions, see Margaret Rose's *The Post-Modern and the Post-Industrial: A Critical Analysis*.

13. It seems abundantly clear that certainly most, if not all, definitions of postmodernism are constructed for particular purposes, to conform to specific agendas, to illustrate certain problems, or to begin to enunciate particular possible solutions. It is not surprising, therefore, to find repeatedly, attached to the beginning of definitions, phrases like Hutcheon's "*I* have offered *my* definition here at the start ..."; "I would like to begin by arguing that, *for me*, postmodernism is ..." (*Politics* 15, *Poetics* 3); or to find John Barth's remarks prefaced with a survey of definitions, but proceeding to what appears to be the "real thing" with "*In my view* ..." ("Replenishment" 201); or to find Terry Eagleton's "*In my view...*" (72) as a context for what follows.

14. It is not at all clear what is being "subverted" in Hutcheon's postmodernism. Surely a fiction that insists that it be recognised only ever as (precisely) fiction, that claims, furthermore, that *all* discourses are only ever fictional, that announces that fiction (and therefore discourse generally) is (only, ever, always) story-telling, and that then celebrates story-telling for its own sake, is not "subverting" fiction, but doing quite the opposite. If anything is being subverted in that scenario, in fact, it is perhaps traditional notions of *history* as a stable, truth- and reality-based discourse, and not fiction at all. This easy use of "subversive" as a laudatory characteristic of postmodernist literature is, I think, a serious problem with many of Hutcheon's formulations; it is a use also shared by Alison Lee (see her *Realism and Power: Postmodern British Fiction*), among others.

15. Neil Postman goes so far as to argue that "... like the printing press, television is nothing less than a philosophy of rhetoric," and that that

"philosophy of rhetoric" has essentially taken over our public discourse: "We are now a culture whose information, ideas and epistemology are given form by television, not by the printed word" (17, 28). And the most significant attribute of the rhetoric of television, by far, is irony. Mark Crispin Miller, while conceding "TV's pure and total irony," also notes that "that inert, ironic watchfulness which TV reinforces in its audience is itself conducive to consumption. As we watch, struggling inwardly to avoid resembling anyone who might stand out as pre- or non- or antitelevisual, we are already trying to live up, or down, to the same standard of acceptability that TV's ads and shows define collectively: the standard that requires the desperate use of all those goods and services that TV proffers..." (16, 327). And Bill McKibben argues that "TV retains its power in part because it has trained us not to take it seriously"; "it is designed to make you say, 'Yeah, I'm hip. I'm just sitting here, but I'm under no illusion that I'm not. I'm not pretending to be something deep. I'm a couch potato'" ("Reflections" 78). All three would agree, I think, that any final construction of meaning in television messages is up to the reader/viewer. Indeed, the Neilsen television ratings are conducted in a way that precisely avoids any record of actual interpretations of the programming, or any measure of retention of content of the programming: the ratings are purely a measure of whether the television set was turned on and tuned to a particular channel, and when, and for how long. It is the viewers' own watchfulness, in fact, that is the product that is actually sold on television, and that is a further irony: television viewers, despite their watchfulness about the products that are advertised so extensively on television, are *themselves* the product that is really being sold.

16. I shall leave for elsewhere a discussion of whether a postmodernist economics is connected, really, to what Charles Newman calls "climax inflation," or whether it is more intimately tied to an ultra-conservative "neo-feudalism."

Chapter 3

1. See, for instance, the work of Teresa Ebert. Ebert aspires to "supersede" "*ludic feminism*: a feminism that is founded upon poststructuralist assumptions about linguistic play, difference, and the

priority of discourse" (3). Ebert aspires to "a revolutionary and dialectical feminism" (3), which feminism would stand in contrast to ludic postmodernism, which appears in Ebert's work as fundamentally empty and politically bankrupt.

2. The others are the "material" drive, composed of concerns such as those for and about "reality" and "the preservation of life"; and the "formal" drive, which is concerned about "the necessity of things," for instance, and the "maintenance of dignity" (105).

3. This attitude towards play and the aura of danger surrounding any writings by Aristotle on play and humour is, of course, one of the basic plot structures in Eco's *The Name of the Rose*.

4. I do not wish to imply, by the way, that literary theory and philosophy are the only spheres of activity that have embraced play as centrally important to an understanding of our current reality.

For instance, there are numerous studies of the psychological importance of play to humans. As an example, Susanna Miller claims in her *The Psychology of Play* that for human children, "Play is ... controlled by collective discipline and codes of honour so that games with rules replace the individual symbolic make-believe of the earlier stage. Although games with rules are socially 'adapted' and survive into adulthood, they still show assimilation rather than adaptation to reality. The rules of the game legitimize the individual's satisfaction in his sensory-motor and intellectual achievement and victory over others, but are not equivalent to intelligent adaptation to reality" (55–56). Miller's notions of the importance of play to humans is rooted in the importance of play to children.

Richard Williams agrees; he writes that "... art is one of the extensions into adult life of the child's world of play" (iii). This is, of course, not to claim that art is childish, merely that it is playful, circular, and therefore always childlike, fresh, and new: "Play and games have this circularity about them; they thrive on repetition. Games are not exhausted in being continually replayed.... The powers of literature and play reside in this untiring circularity of beginning and ending, for the pleasure comes not from the accomplishment but from the actual reading and playing themselves" (240).

And Michael Ellis writes in his *Why People Play* that "At the beginning, each activity involves novelty. The unfamiliarity of the task itself is sufficient to allow the elevation of arousal. As the task is learned then the uncertainty, and hence the impact of the activity, is maintained by the elevation of standards of performance. The probability of achievement of a goal response is maintained at some level so that the uncertainty of its achievement is optimally arousing to the individual. The establishment of the difficulty of the goal and likelihood of its achievement can usually be varied by the player and boils down to the levels of risk of success and failure the individual will take" (142).

Finally, James Hans writes that "Play has come into the foreground of our thought at the same time as our conception of a centered, continuous world has been called into question" (ix). Play is for Hans, therefore, primarily a psychological response to an incoherent reality.

5. *The Play of Man* is actually a companion piece to Groos's first book, *The Play of Animals*; and animals do not have "culture," though they play.

6. Ehrmann writes, "In writing about play, it is impossible to ignore Huizinga's book, *Homo Ludens*" ("Homo Ludens Revisited" 31).

7. See, for example, Mary Warnock, who writes, "in talking about images we are talking not only about a class of *things which represent*, but about a species of thinking" (159). In her stance, Warnock is influenced not only by Huizinga, but by Sartre, for instance, who insists that "an image is *a certain type of consciousness*. An image is an act, not some thing. An image is a consciousness *of* some thing" (146).

The problem can also be approached from a slightly different angle. In Edward Casey's work, play is part of the imagination, rather than the imagination's being part of play, as it is in Huizinga. Casey writes, "In imagining, the mind moves in many ways. Imagination multiplies mentation and is its freest form of movement. It is mind in its polymorphic profusion. It is also mind in the process of self-completion, and as such includes an element of self-enchantment..." (233). Play may be free, and spontaneous, but it is not by itself imagination. Casey is not interested, however, in having his theory become one of "those theories which define imagination as play" (5). Instead, he writes, "... spontaneity and controlledness [emerge] as essential traits of imagination in both of its basic intentional phases.... the two essential traits not only contrast with but also *complement* each other:

what is lacking in spontaneity is found precisely in controlledness, and vice versa" (63).

See also Kendall Walton's discussions of painting and fiction in terms of "the imaginary" and "make-believe": "The fictional truth value of a proposition is always relative to some particular game of make-believe, daydream, or other suitable fictional context" (287–88).

8. To witness Derrida at his relentlessly playful best, see, for instance, the "debate" with John Searle that appears in *Limited Inc.* Following is a typical part of that one-sided "exchange" (Searle apparently refused to allow the essay upon which Derrida's response is based to appear in Derrida's book): "Let's be serious. Faced with this speech act ('let's be serious'), readers may perhaps feel authorized in believing that the presumed signatory of this text is only now beginning to be serious, only now committing himself to a philosophical discussion worthy of the name, and is thus admitting that what he has previously been engaged in was something entirely different. But let's be serious. Why am I having such difficulty being serious..." (34).

For a lucid use of Derrida's theories, see Gregory Ulmer's discussion of a deconstructionist pedagogy in *Applied Grammatology*.

9. See Douglas Hofstadter's treatments of decision theory in the form of Prisoner's Dilemma problems and so-called "Luring Lotteries" in his *Metamagical Themas* 715–55.

10. And it appears that the strategy is enormously successful: Suits calculates 1,023 ways in which a detective story could be a game ("The Detective Story" 219).

11. I discuss the literature/game/puzzle analogy at more length in my "Playing the Games we Read."

12. I shall have more to say about *The Universal Baseball Association* later; the point, for now, is that the novel is not the table-top dice game.

13. I investigated the contest. My searches reached to Margaret Drabble herself. She does not remember who wrote which chapters; she does, however, recall that nobody won the contest.

14. This is reminiscent of a kind of hyperfiction, wherein previously written works are extensively annotated; the information in the notes is accessible to curious readers when they "click" on key words and phrases. Thus the reader who clicked on "April," say, at the beginning of a hypertext version of Eliot's *The Waste Land* would be directed to Chaucer's *Canterbury Tales*, and possibly to an extended discussion of irony, on one hand, and intertextuality and allusion, on the other.

15. And for an extra twist on this technique, see Mark Henshaw's *Out of the Line of Fire*, which begins with a citation followed by an elaborate discussion of the beginning of the Calvino novel, and the implications those two strategies have for Henshaw's novel.

16. *In Palamedes' Shadow* 80 n.10; see also 264, and Suits, *The Grasshopper* 152.

17. See, for instance, the multi-faceted discussions of Lewis Carroll's *Alice* books by Gilles Deleuze: "Such a game—without rules, with neither winner nor loser, without responsibility, a game of innocence, a caucus-race, in which skill and chance are no longer distinguishable—seems to have no reality" (*The Logic of Sense* 60). One might suspect that Suits would claim that Alice's caucus race is not, in fact, a game at all: games do not necessarily need winners or losers; some games are precisely an escape from responsibility; and there are games both of pure skill and of pure chance. However, Suits would also claim (and Wilson would agree) that there can be no game if there are no rules. Without rules, Alice's caucus race is something other than a game. As it turns out, whatever else it is, the caucus race is a satire on elections; if there is a game here, it is a word game (inasmuch as elaborate puns are games) centred on the word "race."

18. As I mentioned earlier, it is possible to invent rules and make a game that is based in part on geometry, without, in that process, making (or even implying to make) geometry itself into a game. And just as the addition of rules can make games that operate inside these convention-laden and convention-defined activities, chess could be studied in other terms and dimensions, most of them strictly conventional: one can imagine a study (though it might be difficult to imagine an interesting or useful study that

does this) that maps a style of play with a player's shoe (or hat?) size (or, somewhat more seriously, her social class, gender, or race, for instance).

19. That chess does depend to some extent on convention can be appreciated by trying to imagine the world chess championship being played with live "chess pieces" (that is, people dressed as chess pieces) on a checked tile floor in a downtown mall.

20. This is a comparison that Wilson mentions, and then drops, as less useful to him regarding rules and conventions than the more obviously conventional "pastoral."

21. The same cannot be said of conventions. As Wilson has noted: "Take away the flutes, the idealized setting, even the sheep, and there may still be a pastoral effect" (*In Palamedes' Shadow* 95).

22. M.H. Abrams, in his *A Glossary of Literary Terms*, defines the sonnet as "A lyric poem consisting of a single stanza of fourteen iambic pentameter lines linked by an intricate rhyme scheme."

23. Imagine trying to convince someone that a particular poem is a sonnet, even though the poem in question is written in trochaic hexameter and arranged in couplets. It is, runs the argument, a lyric poem about secular love that is a single stanza and fourteen lines in length. This poetic form clearly ignores, however, some of the important constitutive rules of sonnets in favour, in part, of purely conventional aspects of many sonnets (in this case, for instance, the convention of the sonnet's being about secular love).

24. Indeed, Shakespeare's sonnets might have been disqualified from sonnet-hood because of the variations to the Petrarchan rhyme scheme that his sonnets display. While the rhyme scheme does seem to be an important part of what makes a sonnet a sonnet, the transformations here seem to have been accommodated as "innovations": the rules have changed slightly, and a new form has been invented.

25. As in the world of film, *Forrest Gump* becomes postmodernism in my reading of it, and a rather sentimental love story otherwise: see Ch. 2, n.2, above; or, more fully, my "Like a Box of Chocolates."

Chapter 4

1. Barth terms the products of Friday afternoons spent writing nonfiction "Friday pieces." The collection in which they appear, therefore, is *The Friday Book*.

2. See Barth's "Don't Count on It: A Note on the Number of *The 1001 Nights*" (*The Friday Book* 258–81), for Barth's remarks on "Scheherazade again, examined intimately indeed. It seems a fit note to end the book upon" (258). That essay does the mathematics necessary to make 1001 nights make sense as the number of Scheherazade's "tellings," in more than a metaphoric sense.

3. Barth calls Shahryar's demand/promise/threat of death to the young women of his kingdom, including Scheherazade, a "primordial publish-or-perish ultimatum" ("Tales within Tales within Tales," *The Friday Book* 220).

4. I wonder whether, in order to be technically correct, I should not shift all the double quotation marks to singles, and singles to doubles, and add my own set of quotation marks around the whole thing....

In any case, this climax is reached, and quoted, in *Chimera*, where the pun between sexual and narrative climax is made explicit: "... one might go beyond the usual tales-within-tales-within-tales-within-tales which our Genie had found a few instances of in that literary treasure-house he hoped one day to add to, and conceive a series of, say, *seven* concentric stories-within-stories, so arranged that the climax of the innermost would precipitate that of the next tale out, and that of the next, et cetera, like a string of firecrackers or the chains of orgasms that Shahryar could sometimes set my sister catenating" (24).

5. This observation also appears in "More Troll than Cabbage: Introduction for Tape-and-Live-Voice Performances from the Series *Lost in the Funhouse*," reprinted in *The Friday Book* 77–79, at 79.

6. See the discussion of Poe's story "The Thousand-and-Second Tale of Scheherazade," below.

7. "As I have said in interviews, I know the present only through the television screen, whereas I have a direct knowledge of the Middle Ages" (Eco, *Postscript* 14).

8. I have made use of three major translations of *The Book of the Thousand Nights and a Night*. The first is Richard F. Burton's; the Sindbad cycle appears in that edition in vol. 6. The second is J.C. Mardrus and Powys Mathers's; the Sindbad stories are in vol. 2 in that edition. The third is Edward William Lane's; the Sindbad stories appear there in vol. 3.

Although Simon relishes reading as a child "a volume of tales from *The Arabian Nights*, unexpurgated" (87)—that is, excerpts from the Burton translation which he got from Daisy's mother—the adult Baylor "finds himself reimmersed" specifically in the Mardrus and Mathers translation of *The Thousand and One Nights* (324). Unless otherwise noted, therefore, all quoted references will be to the Mardrus & Mathers translation.

9. Burton calls him "Sindbad the Landsman" (1)—a name which would in some ways be an even better match for the character in Barth's tale (that is, an even better mirror image), except Barth endows the Porter with a passion for sailing, like Sindbad the Sailor's. This passion allows him to bring the various "voyages" of the twinned Sindbads slowly into line with each other.

10. See the first section of "Some Reasons Why...," subtitled "Twins" (*The Friday Book* 1–3).

11. Barth notes that

> Aesthetic realism ... is any set of artistic conventions felt by people on a particular level of a particular culture at a particular period to be literally imitative of their imagination of the actual world. It goes without saying that one generation's or culture's realism is another's patent artifice.... As for aesthetic irrealism—fantasy of the sort addressed by international conferences on the fantastic—it must consist of any set of artistic principles or devices, conventional or otherwise, felt by people on a particular level of a particular culture at a particular

time to be enjoyable and/or significant though understood to be *not* literally imitative of their imagination of the actual world" ("Tales Within Tales Within Tales," *The Friday Book* 222).

12. Edward W. Said's work on the tensions between the Western world and Islam is very informative; see, in particular, *Orientalism* and *Covering Islam*.

13. An excellent example of this is Burton's note on the roc, or rukh; "The word is Persian," says Burton, and has "many meanings" (see 6: 16–17 n.1). Most of Burton's notes are premised on the assumption that "man remembers and combines but does not create" (ibid.); in other words, good research can always unearth the truth behind the fiction.

14. See his autobiographical account, *Sailing Alone Around the World and Voyage of the Liberdade*.

15. "I wondered where the oil rigs were, and the supertankers and other large shipping bound into and out of the Persian Gulf.... Where were the diesel freighters, the motorized trawlers, the naval patrol craft?" (405)

16. On the use of the sexual metaphor to describe the storyteller and the audience in literature, see Marie Maclean's *Narrative as Performance*, especially the section on "The erotic model: the teller and the hearer" (19ff.).

Chapter 5

1. Throughout this chapter, I will generally use the term "Chippewa." Nanapush uses "Anishinabe" to designate his people in *Tracks*; Hodge suggests that Chippewa, Pauline's term of choice, is a "popular adaptation of Ojibway" (277). Baraga uses the English transliteration "Otchipwe" in his dictionary. In any case, unless it promises to cause confusion, I have not altered sources, whether they have chosen to use "Ojibwa," "Chippewa," or "Anishinabe." A relatively complete and satisfying discussion of the differences among these names is to be found in Gerald Vizenor's *Wordarrows*:

> Indian reservations, or federal enclaves, were created with specific geographical and historical place-names, but the tribal people who were contained there had several names. For example, there are Anishinabe or Anishinabeg and Chippewa or Chippeway and Ojibwa or Ojibway all living at White Earth and Leetch Lake and other reservations in Minnesota and Wisconsin and on tribal reserves in Canada. These are not separate tribes or languages but the same people with several names. In the language of the tribal past the families of one of the largest woodland tribes spoke of themselves as the Anishinabe, singular, and Anishinabeg, plural, until they were named Chippewa and Ojibwa in the English language. Before white contact the woodland tribes used the collective name Anishinabeg in general and specific references to themselves and others who spoke the same language.... The collective names for the tribes were not abstract concepts of nationalism, as the expressions "Chippewa Indians" or the "Chippewa Indian Nation" seem to suggest. Most tribes have several names, or at least two names, one which is spoken in the oral tradition in a tribal language and another which is printed and defined in English language lexicons.... (x–xi)

2. The collection edited by Calvin Martin, *The American Indian and the Problem of History*, addresses gaps and intersections between Native American conceptions of the importance of storytelling and postmodernist notions of the textuality of all discourse in provocative ways.

3. It is worth restating the often-cited biographical information that Erdrich herself is of mixed race. She is half Chippewa and half German-American, and notes that "being Indian is something [she is] terribly proud of" (White and Burnside 165). She belongs to the Turtle Mountain Chippewa band, as did her grandfather, Pat Gourneau (see "Where I Ought to Be" 24).

4. "Our family of nine lived on the very edge of town in a house that belonged to the Government and was rented to employees of the Bureau of Indian Affairs boarding school, where both of my parents worked..." ("Where I Ought to Be" 24).

5. See Danziger, throughout, but esp. 97ff. On the allotment of land to the Turtle Mountain Chippewa Band in particular, see Gregory Camp's "Working Out Their Own Salvation."

6. For documentation of the various land claims that were made by the Chippewa against the United States Government, see the various volumes in the Garland series on the Indian Claims Commission, in the bibliography under Horr, ed., in the separate entries under the seven volumes that relate specifically to the Chippewa.

Ralph Barney writes:

> The land problem was confounded by different concepts inherent in the nature of the disparate cultures. The culture of the Europeans who discovered and later settled this continent was basically legalistic, particularly where land was concerned. Land was the subject of "ownership" either by the monarch or his subjects, and "titles" were the capstone of such ownership. "Ownership" in the sense of a legal right was unknown to the Indian. As Justice Black said in *Shoshone Indians v. United States*, 324 U.S. 335, 357 (1945):
>
> > ... Ownership meant no more to them than to roam the land as a great common, and to possess and enjoy it in the same way that they possessed and enjoyed sunlight and the west wind and the feel of spring in the air. Acquisitiveness, which develops a law of real property, is an accomplishment only of the "civilized."
>
> When the Europeans "discovered" the North American continent they found it inhabited by the Indians and the question of their rights aroused a great moral debate. Charles V of Spain sought the advice of the theologian Franciscus de Victoria, primary professor of sacred theology in the University of Salamanca, who suggested that since the aborigines "were true owners, before the Spaniards came among them, both from the public and private point of view," they should be treated with to secure cessions of their lands. This view obviously could not prevail if the European monarchs owned the land and could parcel it out to their subjects.
>
> The matter came to a head in 1823 when Chief Justice John Marshall decided the famous case of *Johnson v. McIntosh*, 21 U.S. (Wheat.) 453. In 1775 the Piankeshaw Indians had sold a tract of land to various individuals. However, in 1881, the United States sold and patented the same land to William McIntosh. Thus the contest was which deed was valid. From a long and detailed examination of the history of Indian relations in this country, Chief Justice Marshall concluded that the legal title was in the United States Government and that the tribes had no right to sell and convey the land (at least, without governmental consent). (14–15)

7. See the discussions of "family" and "clan" in Hodge; in addition, Danziger writes: "Fellow clansmen ... regarded one another as close kinsmen" (11).

8. Claude Lévi-Strauss writes,

> A kinship system does not consist in the objective ties of descent or consanguinity between individuals. It exists only in human consciousness; it is an arbitrary system of representations, not the spontaneous development of a real situation.... kinship is allowed to establish and perpetuate itself only through specific forms of marriage. (50–51)

9. See also Danziger, esp. 97–105. In addition, see Beaulieu; Camp; Murray; and Youngbear-Tibbetts.

10. Dean Flower's review accuses this novel of being Erdrich's worst effort precisely because "it lapses too easily into the mythic, i.e., the anonymous.... such words [as Nanapush's claim to have 'guided the last buffalo hunt' (2)] ... have been worn thin and parodied by Hollywood scriptwriters" (136–37). Flower's review, I think, misses the mark.

11. Weil writes of the White Earth, Minnesota, reservation (a reservation with historical circumstances exceedingly like the ones that obtain in *Tracks*, and like the ones that obtained in Erdrich's home reservation when she was a child—but one whose history has been very well documented) that "although a few Ojibwa, particularly mixed-bloods on the western part of the reservation, had succeeded as farmers, most of the reservation's population had failed in agriculture. This failure was not surprising, since much of White Earth was not suited for farming, and the allottees had no significant help in learning agriculture" (79–80).

12. See Holly Youngbear-Tibbetts on what she calls the "theft of White Earth," a dispute that began at least as long ago as 1920, and "divides the community today [1991]" (96). She writes that "Allotment and the related timber scandals accomplished the nearly complete dispossession of the White Earth Anishinaabeg, and the only voices raised in protest were those of the resident full-bloods who dared to defy Indian agent Simon Michelet" (97).

13. Michael Dorris, Erdrich's late husband and collaborator.

14. The spelling of the name changes frequently, in part because of differences in dialect, and in part because of difficulties with transliteration into English. The character stays identifiably the same, however, and remains significant. Paredes notes that "Nanabushu is an important symbol of traditional cultural identity among contemporary Chippewa" (379). William Jones's book is essentially a collection of "Nanabushu Tales." The collection includes accounts of his birth, the theft of fire, his marriage, and so on. The collection by Bloomfield and Nichols includes examples of oral transcriptions of narratives involving "Nenabush." In particular, it tells the popular story of Nenabush and the ducks (19–23).

In any case, to pretend that Nanapush is anything other than grandiose in a novel like this one would be roughly equivalent to a self-consciously Christian novel with a narrator named Jesus Christ who pretends to be nothing more than "a guy." The novel risks the kind of parody that the reviewers warn against, in fact, if it does not mythologize enough.

15. Readers of Erdrich's other fiction know that even the stories of Nanapush are not strong enough to accomplish their purpose: we know from *Love Medicine* that Lulu does in fact marry into the Morrissey clan: "[Nanapush] never spoke against Moses the way he later would speak against Morrissey, the first of my law-married husbands, the first mistake" (75).

16. Danziger suggests that the majority of deaths from smallpox happened before the turn of the century (120).

17. See Warren 29–33.

18. On 3 March 1871,

> the legal fiction of recognizing the tribes as independent nations with which the United States could enter into solemn treaties was finally set aside after it had continued for nearly a century. The effect of this act was to bring under the immediate control of Congress the relationship of the Government with the Indians and to reduce to simple agreements what had before been accomplished by treaties as with a foreign power. (Hodge 2: 803)

19. Critics, of course, have not always appreciated this aspect of the novel; scc Flowcr's rcvicw, for instance, for an example of this point missed.

20. Once again, Pauline's religious experience is a skewed version of her traditional religion; what she experiences is nearly a parody of the traditional Chippewa experience, which is described by Hilger this way:

> ... it was the duty of the Ojibwa child to gain an identity beyond its borrowed one, and thereby become a functioning member of society. This was accomplished chiefly through fasts for visions.... The vision was the cornerstone of the individual's religious life, the establishment of a personal relationship with a manito. The relation formed the person's character. (122)

21. That is, in Chippewa, "us."

Chapter 6

1. For a detailed discussion of *Voice of Fire* and Newman more generally, see Harold Rosenberg's book *Barnett Newman* and its bibliography; for a more detailed discussion of the controversy surrounding the National Gallery's purchase of the painting, see Phyllis C. Yaffe's article "Barnett Newman's *Voice of Fire*: The Politics of Art Acquisition."

2. See John Enck's interview with John Hawkes, where Hawkes makes his often-quoted pronouncement that "I began to write fiction on the assumption that the true enemies of the novel were plot, character, setting, and theme, and having once abandoned these familiar ways of thinking about fiction, totality of vision or structure was really all that remained" (149).

3. James Morris writes: "In Venice the past and present are curiously interwoven, as in the minds of very old ladies, who ... sometimes complain petulantly about the ill-treatment of cab-horses" (Morris 115). See also Tony Tanner's book *Venice Desired*, in which he writes of carnivals and their implications, especially as they apply to Venice:

> In general we may say of carnival and masquerade that to some degree or other (and we may be dealing with vestigial traces or the slightest of hints) they involve intimations of the inversion or collapse of hierarchy; the destabilization of genre and perhaps gender; a promiscuous mingling of both classes and sexes; a suspension or failure of habitual taxonomies; an experience of the provisionality and fictionality of customary classificatory systems; possibilities of metamorphosis; a liberation from conventions; an undermining of the dualities and binary oppositions on which culture is founded; category collapse, and perhaps just the faintest glimpse of Chaos and old Night. Masking—which can release inhibition while concealing intention, intimate unknowable secrets or depthless enigmas and indicate what Bakhtin calls the rejection of "conformity to oneself"—necessarily raises questions of identity on all levels. Who are you? What are you? (Tanner 41)

4. On labyrinths, their use, their definitions, and their history, see W.H. Matthews, who writes:

> Both words [maze and labyrinth] have come to signify a complex path of some kind, but when we press for a closer definition we encounter difficulties. We cannot, for instance, say that it is "a tortuous branched path designed to baffle or deceive those who attempt to find the goal to which it leads," for, though that description holds good in some cases, it ignores the many cases in which there is only one path, without branches, and therefore no intent to baffle or mislead, and others again in which there is no definite "goal." We cannot say that it is a winding path "bounded by walls or hedges," for in many instances there are neither walls nor hedges. One of the most famous labyrinths, for example, consisted chiefly of a vast and complicated series of rooms and columns. (2)

On the labyrinth as and in literature, of course, none can match Borges.

5. Coover writes:

> The aged professor ... muses [on] ... his once-upon-a-time passage through Filmland, where the two concepts in question—reality, illusion—were truly inseparable: even he could no longer tell them apart, and so he nearly lost his way again. Finally they gave the role to a blond ingenue who looked like a highschool cheerleader from Iowa dressed up for the junior prom. She wore lipstick and blue eyeshade and

> plucked her eyebrows. Her complexion was nice, though I happen to know she had pimples back where her swimsuit covered them. And she refused to dye her hair blue, so they put her in a kind of slinky blue night-gown and shortened her name to the Blue Fairy. (91–92)

6. Which, in turn, is connected to the Italian tradition that plays such a significant role in the story: "the marionette theatre ... the last vestige of the *commédia dell'arte*" (Duchartre 120).

7. It is not surprising that the fox hunt should figure as prominently as it does in a novel about a painting of a famous fox hunter. The hunt frames and bounds large portions of the novel; it has an interesting pedigree with regard to *Whistlejacket*. See, for example, the manual of foxhunting by John Hawkes, esq., *The Meynellian Science*.

Select Bibliography

Abish, Walter. *Alphabetical Africa*. New York: New Directions, 1974.

Abrams, M.H. *A Glossary of Literary Terms*. 6th ed. Orlando, FL: Harcourt Brace Jovanovich, 1993.

Allen, Paula Gunn. *The Sacred Hoop: Recovering the Feminine in American Indian Traditions*. Boston: Beacon Press, 1992.

———. "Bringing Home the Fact: Tradition and Continuity in the Imagination." Swann and Krupat. 563–79.

Apple, Max. "Post-Modernism." *Free Agents*. New York: Harper and Row, 1984.

Arac, Jonathan. Introduction. Arac, ed. ix–xliii.

Arac, Jonathan, ed. *Postmodernism and Politics*. Minneapolis: U of Minnesota P, 1986.

Baraga, R.R. Bishop. *A Dictionary of the Otchipwe Language, Explained in English: Part II, Otchipwe-English*. Montreal: Beauchemin and Valois, 1880.

Barney, Ralph. "The Indian Claims Commission." Horr vol. 1: 13–16.

Barnouw, Victor. *Wisconsin Chippewa Myths and Tales and Their Relation to Chippewa Life*. Madison, WI: U of Wisconsin P, 1977.

Barth, John. *The Last Voyage of Somebody the Sailor*. Boston: Little, Brown, 1991.

———. *The Friday Book: Essays and Other Non-Fiction*. New York: Putnam's, 1984.

———. *Chimera*. New York: Random House, 1972.

———. *Lost in the Funhouse: Fiction for Print, Tape, Live Voice*. Garden City, NY: Doubleday, 1968.

Barthelme, Donald. *Snow White*. New York: Bantam, 1968.

Barthes, Roland. *A Lover's Discourse*. Trans. R. Howard. New York: Hill and Wang, 1978.

———. "The Death of the Author." *Image–Music–Text*. Trans. Stephen Heath. New York: Hill and Wang, 1977. 142–48.

———. *The Pleasure of the Text*. Trans. R. Miller. New York: Hill and Wang, 1975.

———. *S/Z, An Essay*. Trans. R. Miller. New York: Hill and Wang, 1974.

———. *Mythologies*. Trans. A. Lavers. London: Granada, 1973.

Baudrillard, Jean. *Fatal Strategies*. Ed. Jim Fleming. Trans. Philip Beitchman and W.G.J. Niesluchowski. New York: Semiotext(e), 1990.

———. *The Ecstasy of Communication*. Trans. B. & C. Schutze. New York: Semiotext(e), 1988.

———. "Simulacra and Simulations." *Jean Baudrillard: Selected Writings*. Ed. Mark Poster. Cambridge: Polity Press, 1988. 166–84.

———. *The Evil Demon of Images*. Trans. Paul Patton and Paul Foss. Sydney, Australia: Power Institute of Fine Arts, 1987.

Beaulieu, David L. "Curly Hair and Big Feet: Physical Anthropology and the Implementation of Land Allotment on the White Earth Chippewa Reservation." *American Indian Quarterly* 8.4 (1984): 281–314.

Berman, Neil David. *Playful Fictions and Fictional Players: Game, Sport, and Survival in Contemporary American Fiction*. Port Washington, NY: Kennikat Press, 1981.

Bertens, Hans. *The Idea of the Postmodern: A History*. London: Routledge, 1995.

Black, Max. *Perplexities: Rational Choice, the Prisoner's Dilemma, Metaphor, Poetic Ambiguity, and Other Puzzles*. Ithaca, NY: Cornell UP, 1990.

———. *Models and Metaphors: Studies in Language and Philosophy*. Ithaca, NY: Cornell UP, 1962.

Blied, B.J. "Leopoldinen Stiftung (Leopoldine Society)." *New Catholic Encyclopedia*, vol. 8: 664.

Bloom, Harold, et al. *Deconstruction and Criticism*. New York: Continuum, 1979.

Bloomfield, Leonard and John D. Nichols. *The Dog's Children: Anishinaabe Texts Told by Angeline Williams*. Winnipeg, MB: U of Manitoba P, 1991.

Blow-Up. Dir. Michelangelo Antonioni. MGM, 1966.

Bolter, Jay David. *Writing Space: The Computer, Hypertext, and the History of Writing*. Hillsdale, NJ: Lawrence Erlbaum, 1991.

Borges, Jorge Luis. "Pierre Menard, Author of the Quixote." *Labyrinths: Selected Stories and Other Writings*. Ed. Donald A. Yates and James E. Irby. New York: New Directions, 1964. 36–44.

Borgmann, Albert. *Crossing the Postmodern Divide*. Chicago: U of Chicago P, 1992.

Bove, Paul A. "The Ineluctability of Difference: Scientific Pluralism and Critical Intelligence." Arac, ed. 3–25.

Boyd, Michael. *The Reflexive Novel: Fiction as Critique*. Lewisburg, VA: Bucknell UP, 1983.

Brautigan, Richard. *Trout-Fishing in America*. New York: Dell, 1967.

Bruns, Gerald L. "Cain: Or, The Metaphorical Construction of Cities." *Salmagundi* 74–75 (1987): 70–85.

Burke, Ruth E. *The Games of Poetics: Ludic Criticism and Postmodern Fiction*. New York: Peter Lang, 1994.

Burton, Richard F., trans. and annotations. *The Book of the Thousand Nights and a Night: A Plain and Literal Translation of the Arabian Nights Entertainments*. 10 vols. Privately printed by the Burton Club: N.p., n.d.

Caillois, Roger. *Man, Play, and Games*. Trans. M. Barash. London: Thames and Hudson, 1962.

Calinescu, Matei. *Rereading*. New Haven, CT: Yale UP, 1993.

———. *Five Faces of Modernity: Modernism, Avant-Garde, Decadence, Kitsch, Postmodernism*. Durham, NC: Duke UP, 1987.

Calvino, Italo. *The Castle of Crossed Destinies*. Trans. William Weaver. New York: Harcourt, Brace, Jovanovich, 1977.

Camp, Gregory S. "Working Out Their Own Salvation: The Allotment of Land in Severalty and the Turtle Mountain Chippewa Band, 1870–1920." *American Indian Culture and Research Journal* 14.2 (1990): 19–38.

Caramello, Charles. *Silverless Mirrors: Book, Self and Postmodern American Fiction*. Tallahassee: UP of Florida, 1983.

Casey, Edward S. *Imagining: A Phenomenological Study*. Bloomington: Indiana UP, 1976.

Castillo, Susan Pérez. "Postmodernism, Native American Literature and the Real: The Silko-Erdrich Controversy." *The Massachusetts Review* 32.1 (1991): 285–94.

Cervantes Saavedra, Miguel de. *The First Part of The Life and Adventures of the Renowned Don Quixote de la Mancha*. Trans. Peter Motteux. New York: Random House, 1946.
Chambers, Iain. *Border Dialogues: Journeys in Postmodernity*. London: Routledge, 1990.
Chambers, Ross. "'Narrative' and 'Textual' Functions (with an Example from La Fontaine)." James Phelan, ed. 27–39.
Coleman, Sister Bernard, Ellen Frogner, and Estelle Eich. *Ojibwa Myths and Legends*. Minneapolis, MN: Ross & Haines, 1962.
Collins, Michael and Andreas Papadakis. *Post-Modern Design*. London: Academy Editions, 1989.
Collodi, Carlo. *The Adventures of Pinocchio: Story of a Puppet*. Trans. with Introduction and Notes by Nicolas J. Perella. Berkeley: U of California P, 1991.
Connor, Stephen. *Postmodernist Culture: An Introduction to Theories of the Contemporary*. Oxford: Basil Blackwell, 1989.
Coover, Robert. "Hyperfiction: Novels for the Computer." *The New York Times Book Review*. 29 August 1993: 1+.
———. "The End of Books." *The New York Times Book Review*. 21 June 1992: 1+.
———. *Pinocchio in Venice*. New York: Linden Press, 1991.
———. *Whatever Happened to Gloomy Gus of the Chicago Bears?* New York: Linden Press, 1987.
———. *The Public Burning*. New York: Viking, 1977.
———. *The Universal Baseball Association, Inc. J. Henry Waugh, Prop.* London: Rupert Hart-Davis, 1970.
———. *Pricksongs and Descants*. New York: Dutton, 1969.
———. *The Universal Baseball Association, Inc. J. Henry Waugh, Prop.* New York: Random House, 1968.
———. *The Origin of the Brunists*. New York: Putnam, 1966.
Copway, G. "Kahgegagahbowh." *The Traditional History and Characteristic Sketches of the Ojibway Nation*. 1850. Toronto: Coles, 1972.
Cortázar, Julio. "Blow-Up." *End of the Game and Other Stories*. Trans. P. Blackburn. New York: Pantheon, 1967. 114–31.
Danziger, Edmund Jefferson. *The Chippewas of Lake Superior*. Norman, OK: U of Oklahoma P, 1978.

de Man, Paul. *The Resistance to Theory*. Minneapolis: U of Minnesota P, 1986.

Delaney, John J. *Dictionary of Saints*. Garden City, NY: Doubleday, 1980.

Deleuze, Gilles. *The Logic of Sense*. Trans. Mark Lester. New York: Columbia UP, 1990.

Densmore, Frances. *Chippewa Customs*. Bureau of American Ethnology Bulletin 86. Washington: Government Printing Office, 1929.

Derrida, Jacques. *Limited Inc.* Trans. Samuel Weber and Jeffrey Mehlman. Evanston, IL: Northwestern UP, 1988.

———. *Positions*. Trans. Alan Bass. Chicago: U of Chicago P, 1981.

———. "Living On: Border Lines." Bloom et al. 75–176.

———. *Writing and Difference*. Trans. Alan Bass. Chicago: U of Chicago P, 1978.

———. *Of Grammatology*. Trans. Gayatri Chakravorty Spivak. Baltimore: Johns Hopkins UP, 1977.

Dibble, Julian. "A Rape in Cyberspace." *The Village Voice*. 21 December 1993: 36–42.

Docherty, Thomas, ed. *Postmodernism: A Reader*. New York: Columbia UP, 1993.

Doyle, Arthur Conan. "Silver Blaze." *The Complete Sherlock Holmes*. 2 vols. Toronto: Coles, 1980. 2: 521–36.

Drabble, Margaret and B.S. Johnson, eds. *London Consequences*. London: Greater London Arts Association, 1972.

Drabble, Margaret. Personal letter to Norman Sacuta. 27 March 1995.

Duchartre, Pierre Louis. *The Italian Comedy: The Improvisation Scenarios, Lives, Attributes, Portraits, and Masks of the Illustrious Characters of the Commedia dell'Arte*. Trans. R.T. Weaver. New York: Dover, 1966.

Eagleton, Terry. "Capitalism, Modernism, and Postmodernism." *New Left Review* 152 (1985): 60–73.

Ebert, Teresa L. *Ludic Feminism and After: Postmodernism, Desire, and Labor in Late Capitalism*. Ann Arbor, MI: U of Michigan P, 1996.

Eco, Umberto. *The Limits of Interpretation*. Bloomington: Indiana UP, 1990.

———. *Postscript to* The Name of the Rose. Trans. W. Weaver. New York: Harcourt Brace Jovanovich, 1984.

———. *The Name of the Rose*. Trans. W. Weaver. New York: Harcourt Brace Jovanovich, 1983.

———. *Travels in Hyperreality: Essays*. Trans. W. Weaver. New York: Harcourt Brace Jovanovich, 1983.
Edwards, Brian. *Theories of Play and Postmodernism*. New York: Garland, 1998.
———. "The Player Played: Aspects of Play in Postmodernist Fiction." Diss. U of Melbourne, 1989.
Edwards, Paul, ed. *The Encyclopedia of Philosophy*. New York: Macmillan, 1967.
Ehrmann, Jacques, ed. *Game, Play, Literature. Yale French Studies* 41 (1968).
Ehrmann, Jacques. "The Death of Literature." *New Literary History* 3.1 (1971): 31–47.
———. "Homo Ludens Revisited." Trans. Cathy and Phil Lewis. Ehrmann, ed. 31–57.
Eisenstein, Elizabeth L. *The Printing Press as an Agent of Change*. 2 vols. Cambridge: Cambridge UP, 1979.
Eliade, Mircea. *The Myth of the Eternal Return*. Trans. Willard R. Trask. New York: Bollingen, 1965.
Elkin, Stanley. "In Darkest Hollywood. At the Academy Awards: Voices, Vanity, Vertigo." *Harper's*. December 1989: 51–61.
Ellis, Michael J. *Why People Play*. Englewood Cliffs, NJ: Prentice Hall, 1973.
Enck, John J. "John Hawkes: An Interview." *Wisconsin Studies in Contemporary Literature* 6.2 (1965): 141–55.
Erdrich, Louise. *Love Medicine*. New and Expanded Edition. New York: HarperPerennial, 1993 (1984).
———. *Baptism of Desire*. New York: Harper & Row, 1989.
———. *Tracks*. New York: Henry Holt, 1988.
———. *The Beet Queen*. New York: Henry Holt, 1986.
———. "Where I Ought to Be: A Writer's Sense of Place." *The New York Times Book Review*. 28 July 1985: 1+.
Ewers, John C. *Ethnological Report on the Chippewa Cree Tribe of the Rocky Boy Reservation and the Little Shell Band of Indians*. Horr vol. 6.
Federman, Raymond, ed. *Surfiction: Fiction Now ... and Tomorrow*. 2nd ed., enlarged. Chicago: Swallow Press, 1981.
Fink, Eugen. "The Oasis of Happiness: Toward an Ontology of Play." Trans. Ute and Thomas Saine. Ehrmann, ed. 19–30.

Fish, Stanley. "Why Literary Criticism is Like Virtue." *London Review of Books*. 10 June 1993: 11–16.

———. *Doing What Comes Naturally: Change, Rhetoric, and the Practice of Theory in Literary and Legal Studies*. Durham, NC: Duke UP, 1989.

———. *Is There a Text in This Class?* Cambridge, MA: Harvard UP, 1980.

Fitzgerald, F. Scott. *Tender is the Night*. New York: Scribner's, 1951.

———. *The Great Gatsby*. New York: Scribner's, 1925.

Flower, Dean. Review of *Tracks*. *Hudson Review* 42.1 (1989): 136–37.

Fogel, Stan and Gordon Slethaug. *Understanding John Barth*. Columbia, SC: U of South Carolina P, 1990.

Fogel, Stanley. *The Postmodern University: Essays on the Deconstruction of the Humanities*. Toronto: ECW Press, 1988.

Fokkema, Douwe. *Literary History, Modernism, and Postmodernism*. Amsterdam and Philadelphia: John Benjamins, 1984.

Forrest Gump. Dir. Robert Zemeckis. Paramount, 1994.

Foster, Hal, ed. *The Anti-Aesthetic: Essays on Postmodernism*. Seattle, WA: Bay Press, 1983.

Foster, Hal. *Recodings: Art, Spectacle, Cultural Politics*. Seattle, WA: Bay Press, 1985.

———. "Postmodernism: A Preface." Foster, ed. ix–xvi.

Foucault, Michel. *The Order of Things: An Archaeology of the Human Sciences*. London: Tavistock, 1974.

———. *The Archaeology of Knowledge & The Discourse on Language*. Trans. A.M. Sheridan Smith. New York: Harper Colophon, 1972.

Fowles, John. *The Magus: A Revised Version*. New York: Dell, 1978.

Furlan, W.P. "Pierz, Francis Xavier." *New Catholic Encyclopedia*, vol. 11: 353.

Gadamer, Hans-Georg. *Truth and Method*. 2nd ed. Trans. J. Weinsheimer and D. Marshall. New York: Crossroad, 1989.

Gado, Frank. *First Person: Conversations on Writers and Writing*. Schenectady, NY: Union College Press, 1973.

Garver, Newton. "Rules." P. Edwards, ed., *The Encyclopedia of Philosophy*, vol. 7: 231–33.

Gass, William H. "The Death of the Author." *Salmagundi* 65 (1984): 3–26.

Groos, Karl. *The Play of Man*. Trans. E.L. Baldwin. 1901. New York: Appleton, 1916.

Grossberg, Lawrence, ed. "On Postmodernism and Articulation: An Interview with Stuart Hall." *Journal of Communication Inquiry* 10 (1986): 45–60.

Grossberg, Lawrence. "Putting the Pop Back into Postmodernism." Ross, ed., 167–90.

———. "History, Politics and Postmodernism: Stuart Hall and Cultural Studies." *Journal of Communication Inquiry* 10 (1986): 61–77.

Halliburton, Richard. *The Flying Carpet. The Famous Adventures of Richard Halliburton*. New York: Bobbs-Merrill, 1940.

Hans, James S. *The Play of the World*. Amherst, MA: U of Massachusetts P, 1981.

Harvey, David. *The Condition of Postmodernity*. Cambridge, MA: Blackwell, 1990.

Hassan, Ihab. "Toward a Concept of Postmodernism." *The Postmodern Turn*. Columbus, OH: Ohio State UP, 1987. 84–96. Repr. Docherty. 146–56.

———. "POSTmodernISM: A Paracritical Bibliography." *New Literary History* 3.1 (1971): 5–30.

Hawkes, John, esq. *The Meynellian Science or Fox-Hunting upon System*. 1808. Leicester: Edgar Backus, 1932.

Hawkes, John. *Whistlejacket*. Toronto: Random House, 1988.

———. *Adventures in the Alaskan Skin Trade*. New York: Simon and Schuster, 1985.

Hay, John and Marie MacLean, eds., *Narrative Issues: Special Issue of AUMLA* 74 (1990).

Henshaw, Mark. *Out of the Line of Fire*. London: Hamish Hamilton, 1989.

Hickerson, Harold. *Ethnohistory of Mississippi Bands and Pillager and Winnibigoshish Bands of Chippewa*. Horr vol. 2.

———. *Ethnohistory of Chippewa of Lake Superior*. Horr vol. 3.

———. *Ethnohistory of Chippewa in Central Minnesota*. Horr vol. 4.

Hicks, D. Emily. *Border Writing: The Multidimensional Text*. Minneapolis: U of Minnesota P, 1991.

Hilger, Sister M. Inez. *Chippewa Child Life and Its Cultural Background*. Bureau of American Ethnology Bulletin 146. Washington: Government Printing Office, 1951.

Hobson, Geary, ed. *The Remembered Earth: An Anthology of Contemporary Native American Literature*. Albuquerque, NM: Red Earth Press, 1979.

Hodge, Frederick Webb, ed. *Handbook of American Indians North of Mexico*. 2 vols. 1912. Bureau of American Ethnology Bulletin 30. Washington: Government Printing Office, 1968.

Hofstadter, Douglas R. *Metamagical Themas: Questing for the Essence of Mind and Pattern*. New York: Basic Books, 1985.

Holub, Robert C. *Crossing Borders: Reception Theory, Poststructuralism, Deconstruction*. Madison, WI: U of Wisconsin P, 1992.

Homer. *The Odyssey of Homer*. Trans. S.H. Butcher and A. Lang. New York: Modern Library, 1950.

Horr, David Agee, ed. *American Indian Ethnohistory: North Central and Northeastern Indians: Chippewa Indians*. 7 vols. Garland American Indian Ethnohistory Series. New York: Garland, 1974.

The Holy Bible. Authorized King James Version. Grand Rapids, MI: Zondervan.

Huizinga, Johan. *Homo Ludens: A Study of the Play Element in Culture*. 1950. Boston: Beacon Press, 1955.

Hume, Kathryn. *Fantasy and Mimesis: Responses to Reality in Western Literature*. New York: Methuen, 1984.

Hunter, Jefferson. *Image and Word: The Interaction of Twentieth-Century Photographs and Texts*. Cambridge, MA: Harvard UP, 1987.

Hutcheon, Linda. *The Politics of Postmodernism*. New York: Routledge, 1989.

———. *The Canadian Postmodern: A Study of Contemporary English-Canadian Fiction*. Toronto: Oxford UP, 1988.

———. *A Poetics of Postmodernism: History, Theory, Fiction*. New York: Routledge, 1988.

———. *Narcissistic Narrative: The Metafictional Paradox*. Waterloo, ON: Wilfrid Laurier UP, 1980.

Hyland, Drew A. *The Question of Play*. Lanham, MD: University Press of America, 1984.

Indian Claims Commission. *Commission Findings on the Chippewa Indians*. Horr vol. 7.

Iser, Wolfgang. *The Fictive and the Imaginary: Charting Literary Anthropology*. Baltimore: Johns Hopkins UP, 1993.

———. *The Act of Reading: A Theory of Aesthetic Response*. Baltimore: Johns Hopkins UP, 1978.

Jacobs, Naomi. *The Character of Truth: Historical Figures in Contemporary Fiction*. Carbondale: Southern Illinois UP, 1990.

Jameson, Fredric. *Postmodernism, or, The Cultural Logic of Late Capitalism*. Durham, NC: Duke UP, 1991.

———. "Postmodernism, or the Cultural Logic of Late Capitalism." *New Left Review* 146 (1984): 53–92.

———. "Postmodernism and Consumer Society." Foster, ed. 111–25.

Jencks, Charles. *Post-Modernism: The New Classicism in Art and Architecture*. New York: Rizzoli, 1987.

Johnson, B.S. *Aren't You Rather Young to be Writing Your Memoirs?* London: Hutchinson, 1973.

———. *The Unfortunates*. London: Panther Books in association with Secker and Warburg, 1969.

Jones, William, collector. *Ojibwa Texts*. Ed. Truman Michelson. New York: G.E. Stechert (Publications of the American Ethnographic Society, vol. 7, part 1. Ed. Frank Boas), 1917.

Joyce, Michael. *Afternoon, A Story*. Cambridge, MA: Eastgate Systems, 1987.

Kennedy, William. *Billy Phelan's Greatest Game*. New York: Viking, 1978.

Kinser, Samuel. *Rabelais's Carnival: Text, Context, Metatext*. Berkeley: U of California P, 1990.

Klinkowitz, Jerome. *Structuring the Void: The Struggle for Subject in Contemporary American Fiction*. Durham and London: Duke UP, 1992.

Klotz, Heinrich. *The History of Postmodern Architecture*. Trans. R. Donnell. Cambridge, MA: The MIT Press, 1988.

Knuth, Helen E. *Economic and Historical Background of Northeastern Minnesota Lands: Chippewa Indians of Lake Superior*. Horr vol. 3.

Kristeva, Julia. *Desire in Language: A Semiotic Approach to Literature and Art*. Ed. Leon S. Roudiez. Trans. Thomas Gora, Alice Jardine, and Leon S. Roudiez. New York: Columbia UP, 1980.

Kroker, Arthur, Marilouise Kroker, and David Cook. *Panic Encyclopedia: The Definitive Guide to the Postmodern Scene*. Montreal: New World Perspectives, 1989.

Kroker, Arthur and David Cook. *The Postmodern Scene: Excremental Culture and Hyper-Aesthetics*. Montreal: New World Perspectives, 1987.

Kroker, Arthur. *Spasm*. Montreal: New World Perspectives, 1994.

Landes, Ruth. *Ojibwa Religion and the Midewiwin*. Madison, WI: U of Wisconsin P, 1968.

Landow, George P., ed. *Hyper/Text/Theory*. Baltimore: Johns Hopkins UP, 1994.

Landow, George P. *Hypertext: The Convergence of Contemporary Critical Theory and Technology*. Baltimore: Johns Hopkins UP, 1992.

Lane, Edward William, trans. and annotations. *The Thousand and One Nights, Commonly Called, in England, the Arabian Nights' Entertainments*. 3 vols. London: Chatto and Windus, 1889.

LeClair, Tom and Larry McCaffery, eds. *Anything Can Happen: Interviews with Contemporary American Authors*. Urbana, IL: U of Illinois P, 1988.

Lee, Alison. *Realism and Power: Postmodern British Fiction*. London: Routledge, 1990.

Levi, Sister M. Carolissa. *Chippewa Indians of Yesterday and Today*. New York: Pageant, 1956.

Lévi-Strauss, Claude. *Structural Anthropology*. Trans. C. Jacobson and B.G. Schoepf. New York: Basic Books, 1963.

Lewis, David. *On the Plurality of Worlds*. Oxford: Blackwell, 1986.

———. *Convention: A Philosophical Study*. Cambridge, MA: Harvard UP, 1969.

Lodge, David. *The Modes of Modern Writing: Metaphor, Metonymy, and the Typology of Modern Literature*. 1977. Chicago: U of Chicago P, 1988.

Lyotard, Jean-François. *The Postmodern Condition: A Report on Knowledge*. Trans. Geoff Bennington and Brian Massumi. Minneapolis: U of Minnesota P, 1984.

Maclean, Marie. *Narrative as Performance: The Baudelairean Experiment*. New York: Routledge, 1988.

Maltby, Paul. *Dissident Postmodernists: Barthelme, Coover, Pynchon*. Philadelphia: U of Pennsylvania P, 1991.

Mardrus, J.C. and Powys Mathers, trans. *The Book of the Thousand Nights and One Night*. 4 vols. New York: St. Martin's, 1972.

Maristuen-Rodakowski, Julie. "The Turtle Mountain Reservation in North Dakota: Its History as Depicted in Louise Erdrich's *Love Medicine* and *Beet Queen*." *American Indian Culture and Research Journal* 12.3 (1988): 33–48.

Martin, Calvin, ed. *The American Indian and the Problem of History*. New York: Oxford UP, 1987.

Matthews, W.H. *Mazes and Labyrinths: A General Account of Their History and Developments*. New York: Longmans, Green, 1922.
McCaffery, Larry. *The Metafictional Muse*. Pittsburgh: U of Pittsburgh P, 1982.
McHale, Brian. *Constructing Postmodernism*. London: Routledge, 1992.
———. *Postmodernist Fiction*. New York: Methuen, 1987.
McKibben, Bill. *The Age of Missing Information*. New York: Plume, 1993.
———. "Reflections: What's On?" *The New Yorker*. 9 March 1992: 40–80.
McNamee, Gregory, ed. *Living in Words: Interviews from the Bloomsbury Review 1981–1988*. Portland, OR: Breitenbush Books, 1988.
Melville, Herman. *Moby Dick; or, The Whale*. 1851. New York: Norton, 1976.
Messenger, Christian K. *Sport and the Spirit of Play in American Fiction, Hawthorne to Faulkner*. New York: Columbia UP, 1981.
Miller, Mark Crispin. *Boxed In: The Culture of TV*. Evanston, IL: Northwestern UP, 1988.
Miller, Susanna. *The Psychology of Play*. Harmondsworth: Penguin, 1968.
Momaday, N. Scott. *In the Presence of the Sun*. New York: St. Martin's, 1992.
———. "The Man Made of Words." Hobson. 162–73.
Morison, Samuel Eliot. *The Oxford History of the American People*. New York: Oxford UP, 1965.
Morris, James. *Venice*. London: Faber and Faber, 1960.
Moulthrop, Stuart. *Victory Garden*. Cambridge, MA: Eastgate Systems, 1991.
Murray, Stanley N. "The Turtle Mountain Chippewa, 1882–1905." *North Dakota History* 51.1 (1984): 14–37.
Natural Born Killers. Dir. Oliver Stone. Warner Bros., 1994.
New Catholic Encyclopedia. New York: McGraw-Hill, 1967.
Newman, Charles. *The Post-Modern Aura: The Act of Fiction in an Age of Inflation*. Evanston, IL: Northwestern UP, 1985.
Nichols, Bill. *Blurred Boundaries: Questions of Meaning in Contemporary Culture*. Bloomington: Indiana UP, 1994.
Norris, Christopher. *What's Wrong with Postmodernism: Critical Theory and the Ends of Philosophy*. Baltimore: Johns Hopkins UP, 1990.
Ommundsen, Wenche. *Metafictions? Reflexivity in Contemporary Texts*. Carlton, Victoria: Melbourne UP, 1993.

———. "The Reader in Contemporary Metafiction." Hay and MacLean. 169–84.

Ong, Walter J. *Orality and Literacy: The Technologizing of the Word.* London: Routledge, 1988.

Paredes, J. Anthony, ed. *Anishinabe: Six Studies of Modern Chippewa.* Tallahassee: UP of Florida, 1980.

Peterson, Nancy J. "History, Postmodernism, and Louise Erdrich's *Tracks.*" *PMLA* 109.5 (1994): 982–94.

Phelan, James, ed. *Reading Narrative: Form, Ethics, Ideology.* Columbus, OH: Ohio State UP, 1989.

Phelan, James. *Reading People, Reading Plots: Character, Progression, and the Interpretation of Narrative.* Chicago: U of Chicago P, 1989.

Pinocchio. Dir. Walt Disney. Walt Disney Studios, 1940.

Poe, Edgar Allen. *Tales of Edgar Allan Poe.* New York: Random House, 1944.

Postman, Neil. *Amusing Ourselves to Death: Public Discourse in the Age of Show Business.* New York: Penguin, 1985.

Putz, Manfred and Peter Freese, eds. *Postmodernism in American Literature: A Critical Anthology.* Darmstadt: Thesen-Verlag, 1984.

Pynchon, Thomas. *Vineland.* Boston: Little, Brown, 1990.

Quittner, Josh. "Johnny Manhattan Meets the FurryMuckers." *Wired.* March 1994: 92+.

Racevskis, Karlis. *Postmodernism and the Search for Enlightenment.* Charlottesville, VA: UP of Virginia, 1993.

Rainwater, Catherine. "Reading Between Worlds: Narrativity in the Fiction of Louise Erdrich." *American Literature* 62.3 (1990): 405–22.

Rawls, John. "Two Concepts of Rules." *Philosophical Review* 64 (1955): 3–32.

Readings, Bill and Bennet Schaber, eds. *Postmodernism Across the Ages.* Syracuse, NY: Syracuse UP, 1993.

Rose, Margaret A. *Parody: Ancient, Modern, and Post-Modern.* Cambridge: Cambridge UP, 1993.

———. *The Post-Modern and the Post-Industrial: A Critical Analysis.* Cambridge: Cambridge UP, 1991.

———. *Parody/Meta-Fiction: An Analysis of Parody as a Critical Mirror to the Writing and Reception of Fiction.* London: Croom Helm, 1979.

Rosenberg, Harold. *Barnett Newman.* New York: Abrams, 1978.

Ross, Andrew, ed. *Universal Abandon? The Politics of Postmodernism*. Minneapolis: U of Minnesota P, 1988.

Sartre, Jean-Paul. *Imagination: A Psychological Critique*. Trans. F. Williams. Ann Arbor: U of Michigan P, 1962.

de Saussure, Ferdinand. *Course in General Linguistics*. Ed. C. Bally and A. Sechehaye. Trans. R. Harris. London: Duckworth, 1983.

Scholes, Robert. *Protocols of Reading*. New Haven: Yale UP, 1989.

Schiller, Friedrich. *On the Aesthetic Education of Man in a Series of Letters*. Ed. and trans. E.M. Wilkinson and L.A. Willoughby. Oxford: Clarendon, 1967.

Scott, Steven, ed. "Play, Games, Literature/Jeu, Les Jeux, Littérature." *Canadian Review of Canadian Literature* 23.2 (June 1996): 271–392.

Scott, Steven D. "Like a Box of Chocolates: *Forrest Gump* and Postmodernism." *Literature/Film Quarterly,* Forthcoming.

———. "Playing the Games we Read and Reading the Games we Play." Scott, ed. 373–92.

Searle, John. *Speech Acts: An Essay in the Philosophy of Language*. Cambridge: Cambridge UP, 1969.

Severin, Tim. *The Sindbad Voyage*. London: Hutchinson, 1982.

Sharrock, Floyd W. and Susan R. Sharrock. *History of the Cree Indian Territorial Expansion from the Hudson Bay Area to the Interior Saskatchewan and Missouri Plains*. Horr vol. 6.

Silberman, Robert. "Opening the Text: *Love Medicine* and the Return of the Native American Woman." Vizenor, ed. *Narrative Chance*. 101–20.

Silko, Leslie Marmon. "Here's an Odd Artifact for the Fairy-Tale Shelf." *Studies in American Indian Literature* 10.4 (1986): 177–84. Reprint of a review of *The Beet Queen* which was originally published in *Impact/Albuquerque Journal Magazine*. 7 October 1986: 10–11. Quotations are from the *SAIL* reprint.

Slemon, Stephen. "Modernism's Last Post." *ARIEL* 20 (1989): 3–17.

Slocum, Captain Joshua. *Sailing Alone Around the World and Voyage of the Liberdade*. London: Rupert Hart-Davis, 1955. Introduction by Arthur Ransome.

Sontag, Susan. *On Photography*. New York: Delta, 1977.

———. "Notes on 'Camp.'" *Partisan Review* 31.4 (1964): 515–30.

Spariosu, Mihai I. *God of Many Names: Play, Poetry, and Power in Hellenistic Thought from Homer to Aristotle*. Durham, NC: Duke UP, 1991.

———. *Dionysus Reborn: Play and the Aesthetic Dimension in Modern Philosophical and Scientific Discourse*. Ithaca: Cornell UP, 1989.

Stead, Deborah. "Unlocking the Tale." *New York Times Review of Books*. 2 October 1988: 41.

Steiner, Wendy. "Obscured Objects of Desire." *Times Literary Supplement*. 31 March 1989: 343.

Stephanson, Anders. "A Conversation with Fredric Jameson." Ross 3–30.

Stewart, Susan. *Nonsense: Aspects of Intertextuality in Folklore and Literature*. Baltimore: Johns Hopkins UP, 1979.

Stonehill, Brian. *The Self-Conscious Novel: Artifice in Fiction from Joyce to Pynchon*. Philadelphia: U of Pennsylvania P, 1988.

Stout, David B. *Ethnohistorical Report on the Saginaw Chippewa*. Horr vol. 5.

Stripes, James D. "The Problem(s) of (Anishinaabe) History in the Fiction of Louise Erdrich: Voices and Contexts." *Wicazo Sa Review* 7.2 (1991): 26–33.

Strouse, Jean. Review of *Tracks*. "In the Heart of the Heartland." *New York Times Review of Books*. 2 October 1988: 1+.

Suits, Bernard. *The Grasshopper: Games, Life and Utopia*. Toronto: U of Toronto P, 1990.

———. "The Detective Story: A Case Study of Games in Literature." Wilson, ed. 200–19.

Sukenick, Ronald. *Down and In: Life in the Underground*. New York: Beech Tree Books, 1987.

Swann, Brian and Arnold Krupat, eds. *Recovering the Word: Essays on Native American Literature*. Berkeley: U of California P, 1987.

Szondi, Peter. *On Textual Understanding and Other Essays*. Minneapolis: U of Minnesota P, 1986.

Tanner, Helen Hornbeck. *The Chippewa of Eastern Lower Michigan*. Horr vol. 5.

Tanner, Tony. *Venice Desired*. Oxford: Blackwell, 1992.

Thiher, Allen. *Words in Reflection: Modern Language Theory and Postmodern Fiction*. Chicago: U of Chicago P, 1984.

Tobin, Patricia. *John Barth and the Anxiety of Continuance*. Philadelphia: U of Pennsylvania P, 1992.

Trachtenberg, Stanley, ed. *The Postmodern Moment: A Handbook of Contemporary Innovation in the Arts*. Westport, CT: Greenwood Press, 1985.

Twitchell, James B. *Carnival Culture: The Trashing of Taste in America*. New York: Columbia UP, 1992.

Ulmer, Gregory L. *Applied Grammatology: Post(e)-Pedagogy from Jacques Derrida to Joseph Beuys*. Baltimore: Johns Hopkins, 1985.

———. "The Object of Post-Criticism." Foster, ed. 83–110.

Vecsey, Christopher. "Envision Ourselves Darkly, Imagine Ourselves Richly." Martin, ed. 120–27.

———. *Traditional Ojibwa Religion and Its Historical Changes*. Philadelphia: The American Philosophical Society, 1983.

Velie, Alan. "The Trickster Novel." Vizenor, ed. *Narrative Chance*. 121–39.

Venturi, Robert, Denise Scott Brown, and Steven Izenour. *Learning from Las Vegas: The Forgotten Symbolism of Architectural Form*. 2nd ed. Cambridge, MA: The MIT Press, 1977.

Vizenor, Gerald, ed. *Summer in the Spring: Anishinaabe Lyric Poems and Stories, New Edition*. Norman, OK: U of Oklahoma P, 1993.

———. *Narrative Chance: Postmodern Discourse on Native American Indian Literatures*. Albuquerque: U of New Mexico P, 1989.

Vizenor, Gerald. Introduction. Vizenor, ed. *Narrative Chance*. 3–16.

———. *The People Named the Chippewa: Narrative Histories*. Minneapolis: U of Minnesota P, 1984.

———. *Wordarrows: Indians and Whites in the New Fur Trade*. Minneapolis: U of Minnesota P, 1978.

Vonnegut, Jr., Kurt. *Slaughter-House Five*. New York: Dell, 1971.

Walton, Kendall L. "Pictures and Make-Believe." *The Philosophical Review* 82.3 (1973): 283–319.

Warner, Robert M. and Lois J. Groesbeck. *Historical Report on the Sault Ste. Marie Area*. Horr vol. 5.

Warner, Robert M. *Economic and Historical Report on Northern Michigan*. Horr vol. 5.

Warnock, Mary. *Imagination*. London: Faber and Faber, 1976.

Warren, William W. *History of the Ojibway Nation*. Minneapolis, MN: Ross & Haines, 1957.

Weil, Richard H. "Destroying a Homeland: White Earth, Minnesota." *American Indian Culture and Research Journal* 13.2 (1989): 69–95.

Wheeler-Voegelin, Erminie and Harold Hickerson. *The Red Lake and Pembina Chippewa*. Horr vol. 1.

Wheeler-Voegelin, Erminie. *An Anthropological Report on Indian Use and Occupancy of Northern Michigan*. Horr vol. 5.

White, Sharon and Glenda Burnside. "On Native Ground: An Interview with Louise Erdrich and Michael Dorris." McNamee. 157–69.

Williams, Donna. *Nobody Nowhere*. London: Doubleday, 1992.

Williams, Richard Darby. "*Poeta Ludens*: Explorations in the Theory of Art as Play." Diss. Ohio State U, 1971.

Wilson, R[obert] Rawdon, ed. *Game and the Theories of Game*. Special Issue of *Canadian Review of Comparative Literature* 12.2 (June 1985): 177–370.

Wilson, R[obert] Rawdon. *In Palamedes' Shadow: Explorations in Play, Game, & Narrative Theory*. Boston: Northeastern UP, 1990.

———. "SLIP PAGE: Angela Carter, In/Out/In The Postmodern Nexus." *ARIEL* 20 (1989): 96–114.

Wittgenstein, Ludwig. *Philosophical Investigations*. Trans. G.E.M. Anscombe. Oxford: Basil Blackwell, 1988.

———. *Tractatus Logico-Philosophicus*. Trans. D.F. Pears & B.F. McGuinness. London: Routledge, 1988.

Wollen, Peter. *Readings and Writings: Semiotic Counter-Strategies*. London: Verso, 1982.

Wright, Richard. *Native Son*. New York: Harper & Row, 1966 (1940).

Yaffe, Phyllis C. "Barnett Newman's *Voice of Fire*: The Politics of Art Acquisition." *Queen's Quarterly* 98.4 (1991): 874–91.

Young, Robert, ed. *Untying the Text: A Post-Structuralist Reader*. 1981. London: Routledge & Kegan Paul, 1987.

Youngbear-Tibbetts, Holly. "Without Due Process: The Alienation of Individual Trust Allotments of the White Earth Anishinaabeg." *American Indian Culture and Research Journal* 15.2 (1991): 93–138.

Zavarzadeh, Mas'ud. *The Mythopoeic Reality: The Postwar American Nonfiction Novel*. Chicago: U of Illinois P, 1976.

Zhang, Benzi. "Paradox of Chinese Boxes: Narrative Reflexivity in Postmodern Fiction." Diss. U of Alberta, 1992.

Ziegler, Heide. *John Barth*. Contemporary Writers Series. London: Methuen, 1987.

INDEX